Resource-based Learning in Geography

Charles Seale-Hayne Library
University of Plymouth
(01752) 588 588
LibraryandITenquiries@plymouth.ac.uk

Geography Discipline Network (GDN)

Higher Education Funding Council for England
Fund for the Development of Teaching and Learning

Dissemination of Good Teaching, Learning and Assessment Practices in Geography

Aims and Outputs

The project's aim has been to identify and disseminate good practice in the teaching, learning and assessment of geography at undergraduate and taught postgraduate levels in higher education institutions.

Ten guides have been produced covering a range of methods of delivering and assessing teaching and learning:

- Teaching and Learning Issues and Managing Educational Change in Geography
- Lecturing in Geography
- Small-group Teaching in Geography
- Practicals and Laboratory Work in Geography
- Fieldwork and Dissertations in Geography
- Resource-based Learning in Geography
- Teaching and Learning Geography with Information & Communication Technologies
- Transferable Skills and Work-based Learning in Geography
- Assessment in Geography
- Curriculum Design in Geography

A resource database of effective teaching, learning and assessment practice is available on the World Wide Web, http://www.chelt.ac.uk/gdn, which contains national and international contributions. Further examples of effective practice are invited; details regarding the format of contributions are available on the Web pages. Examples should be sent to the Project Director.

Project Team

Lead site: *Cheltenham & Gloucester College of Higher Education*
Professor Mick Healey; Dr Phil Gravestock; Dr Jacky Birnie; Dr Kris Mason O'Connor

Consortium: *Lancaster University*
Dr Gordon Clark; Terry Wareham
Middlesex University
Ifan Shepherd; Professor Peter Newby
Nene — University College Northampton
Dr Ian Livingstone; Professor Hugh Matthews; Andrew Castley
Oxford Brookes University
Dr Judy Chance; Professor Alan Jenkins
Roehampton Institute London
Professor Vince Gardiner; Vaneeta D'Andrea; Shân Wareing
University College London
Dr Clive Agnew; Professor Lewis Elton
University of Manchester
Professor Michael Bradford; Catherine O'Connell
University of Plymouth
Dr Brian Chalkley; June Harwood

Advisors: Professor Graham Gibbs (*Open University, Milton Keynes*)
Professor Susan Hanson (*Clark University, USA*)
Dr Iain Hay (*Flinders University, Australia*)
Geoff Robinson (*CTI Centre for Geography, Geology and Meteorology, Leicester*)
Professor David Unwin (*Birkbeck College, London*)
Dr John Wakeford (*Lancaster University*)

Further Information

Professor Mick Healey, Project Director   Tel: +44 (0)1242 543364   Email: mhealey@chelt.ac.uk
Dr Phil Gravestock, Project Officer   Tel: +44 (0)1242 543368   Email: pgstock@chelt.ac.uk
Cheltenham & Gloucester College of Higher Education
Francis Close Hall, Swindon Road, Cheltenham, GL50 4AZ, UK   [Fax: +44 (0)1242 532997]

http://www.chelt.ac.uk/gdn

Resource-based Learning in Geography

Mick Healey

Cheltenham & Gloucester College of Higher Education

Series edited by Phil Gravestock and Mick Healey
Cheltenham & Gloucester College of Higher Education

Published by:

Geography Discipline Network (GDN)

Cheltenham & Gloucester College of Higher Education

Francis Close Hall

Swindon Road

Cheltenham

Gloucestershire, UK

GL50 4AZ

Resource-based Learning in Geography

ISBN: 1 86174 029 8

ISSN: 1 86174 023 9

Typeset by Phil Gravestock

Cover design by Kathryn Sharp

Printed by:

Frontier Print and Design Ltd.

Pickwick House

Chosen View Road

Cheltenham

Gloucestershire, UK

Contents

8 How can students and staff be supported? 80

9 Case studies 90

Boxes

Figures

Tables

Editors' preface

This Guide is one of a series of ten produced by the Geography Discipline Network (GDN) as part of a Higher Education Funding Council for England (HEFCE) and Department of Education for Northern Ireland (DENI) Fund for the Development of Teaching and Learning (FDTL) project. The aim of the project is to disseminate good teaching, learning and assessment practices in geography at undergraduate and taught postgraduate levels in higher education institutions.

The Guides have been written primarily for lecturers and instructors of geography and related disciplines in higher education and for educational developers who work with staff and faculty in these disciplines. For a list of the other titles in this series see the information at the beginning of this Guide. Most of the issues discussed are also relevant for teachers in further education and sixth-form colleges in the UK and upper level high school teachers in other countries. A workshop has been designed to go with each of the Guides, except for the first one which provides an overview of the main teaching and learning issues facing geographers and ways of managing educational change. For details of the workshops please contact one of us. The Guides have been designed to be used independently of the workshops.

The GDN Team for this project consists of a group of geography specialists and educational developers from nine old and new universities and colleges (see list at the front of this Guide). Each Guide has been written by one of the institutional teams, usually consisting of a geographer and an educational developer. The teams planned the outline content of the Guides and these were discussed in two workshops. It was agreed that each Guide would contain an overview of good practice for the particular application, case studies including contact names and addresses, and a guide to references and resources. Moreover it was agreed that they would be written in a user-friendly style and structured so that busy lecturers could dip into them to find information and examples relevant to their needs. Within these guidelines the authors were deliberately given the freedom to develop their Guides in their own way. Each of the Guides was refereed by at least four people, including members of the Advisory Panel.

The enthusiasm of some of the authors meant that some Guides developed a life of their own and the final versions were longer than was first planned. Our view is that the material is of a high quality and that the Guides are improved by the additional content. So we saw no point in asking the authors to make major cuts for the sake of uniformity. Equally it is important that the authors of the other Guides are not criticised for keeping within the original recommended length!

Although the project's focus is primarily about disseminating good practice within the UK a deliberate attempt has been made to include examples from other countries, particularly North America and Australasia, and to write the Guides in a way which is relevant to geography staff and faculty in other countries. Some terms in common use in the UK may not be immediately apparent in other countries. For example, in North America for 'lecturer' read 'instructor' or 'professor'; for 'staff' or 'tutor' read 'faculty'; for 'postgraduate' read 'graduate'; and for 'Head of Department' read 'Department Chair'. A 'dissertation' in the

UK refers to a final year undergraduate piece of independent research work, often thought of as the most significant piece of work the students undertake; we use 'thesis' for the Masters/ PhD level piece of work rather than 'dissertation' which is used in North America.

In addition to the Guides and workshops a database of good practice has been established on the World Wide Web (http://www.chelt.ac.uk/gdn). This is a developing international resource to which you are invited to contribute your own examples of interesting teaching, learning and assessment practices which are potentially transferable to other institutions. The resource database has been selected for *The Scout Report for Social Sciences*, which is funded by the National Science Foundation in the United States, and aims to identify only the best Internet resources in the world. The project's Web pages also provide an index and abstracts for the *Journal of Geography in Higher Education*. The full text of several geography educational papers and books are also included.

Running a consortium project involves a large number of people. We would particularly like to thank our many colleagues who provided details of their teaching, learning and assessment practices, many of which appear in the Guides or on the GDN database. We would also like to thank, the Project Advisers, the FDTL Co-ordinators and HEFCE FDTL staff, the leaders of the other FDTL projects, and the staff at Cheltenham and Gloucester College of Higher Education for all their help and advice. We gratefully acknowledge the support of the Conference of Heads of Geography Departments in Higher Education Institutions, the Royal Geographical Society (with the Institute of British Geographers), the Higher Education Study Group and the *Journal of Geography in Higher Education*. Finally we would like to thank the other members of the Project Team, without them this project would not have been possible. Working with them on this project has been one of the highlights of our professional careers.

Phil Gravestock and Mick Healey

Cheltenham

July 1998

All World Wide Web links quoted in this Guide were checked in July 1998. An up-to-date set of hyperlinks is available on the Geography Discipline Network Web pages at:

http://www.chelt.ac.uk/gdn

About the author

Mick Healey

I am an economic geographer, with a particular interest in local economic development. I have written and edited several books, including *Location & Change: Perspectives on Economic Geography* (Oxford University Press, 1990), which I co-authored with Brian Ilbery. In recent years I have also developed an interest in the learning and teaching of geography in higher education.

After studying for my BA and PhD at the University of Sheffield, I taught at Coventry University for 20 years. I then moved to Cheltenham & Gloucester College of Higher Education in 1994 to Head the Department of Geography & Geology. I also taught part-time for the Open University for 12 years. I am currently Professor of Geography in the Geography and Environmental Management Research Unit.

I have learnt much about the wider context of learning and teaching through jointly editing the *Journal of Geography in Higher Education*, acting as a HEFCE Assessor in the Teaching Quality Assessment of Geography, and chairing a group organising an institution-wide debate about the development of teaching and learning in the College. As for resource-based learning, the subject of this Guide, I have helped develop resource packs to integrate the teaching of transferable skills into mainline geography modules and have designed a module based around discussions of chapters from a textbook. I also try to incorporate active learning exercises in my lectures. I recently co-directed the HEFCE Teaching and Learning Technology Programme *GeographyCal* project, developing computer assisted learning packages for introductory geography courses.

Despite my interests in educational development, I remain first and foremost a geographer, who is convinced about the mutually reinforcing benefits of my teaching and research.

I hope that you'll enjoy reading this Guide as much as I have enjoyed preparing and writing it.

1 Introduction

1.1 What is the focus of this Guide?

This Guide is one of ten produced by the Geography Discipline Network (GDN) to identify and disseminate good teaching, learning and assessment practices in geography. Its concern is with resource-based learning.

Although *Resource-based learning (RBL)* may be variously defined, I refer to learning schemes where the emphasis is on the use by students of print and electronic-based learning resources. They vary in length from resource-based exercises used to stimulate short periods of active learning in lectures; through learning packs, designed to replace or supplement a lecture or a block of lectures, and readers, compiled to supplement library resources; to whole courses which are delivered through independent learning packages (Box 1).

Box 1:   Teaching and learning materials can take a variety of forms

These include:

1.   *Straight content*: this may be in the form of a book, WWW sites, indexed and searchable CD-ROMs.

2.   *Readers*: collections of materials, which may also be in electronic formats.

3.   *Tutorials in print*: with in-text questions, activities and commentary.

4.   *Study guides or course guides*: focused on process rather than content, giving advice, activities, glossaries, formulae and examples, often with reference to other kinds of materials.

5.   *Support for class sessions*: materials for students to use in class.

6.   *Tutor guides*: prepared for tutors, with guidelines for using other materials and perhaps answers to problems, additional materials and activities.

7.   *Hybrids*: various combinations of these formats.

Reference:
FDTL National Co-ordination Team (1998a, p.1)

RBL is not new; after all, when we send our students to the library to read a set of references on a lecture topic, this is in one sense a form of resource-based learning. Handouts describing practical and fieldwork exercises are another common form of RBL. Books and journals are likely to remain the most important learning resource for many years to come. However, pressures on library resources, particularly from large modules/units where typically 100 or more students are referred to the same references at the same time, mean that books and journals need to be supplemented by other learning resources. These include study guides, self-learning materials, readers, learning exercises, audio-cassettes, videos, computer-based learning materials, and on-line databanks. Learning packages, which use one or more of these resources, are becoming increasingly common in Higher Education (HE) and all the indications are that their use will increase rapidly to cope with the projected increased numbers of students from more diversified backgrounds and the continuing rise in student-staff ratios (SSRs) (see Sections 3.1.2 and 3.1.3). Hence, as defined here, all staff and students already use RBL to a greater or lesser extent. This Guide is concerned to review the variety of forms which RBL may take and to examine how you may use it to enhance the quality of learning, so that you can make informed judgments about the circumstances under which it is desirable to extend the use of RBL and what you need to do to make its use effective.

This Guide shows that not only can RBL help meet the kind of challenges currently facing geography in HE, but also that, suitably designed, RBL packages and exercises can improve the quality of your students' learning experience and can have advantages for you, including, under specific circumstances, saving you time. RBL, like any other method of learning, may, of course, not work effectively for a whole host of reasons, including inappropriate content, insufficient interactivity, failure to provide induction for students and staff using the materials, absence of feedback, inappropriate assessment, and lack of institutional support. One of the aims of this Guide is to help you plan to minimize the problems and maximize the benefits of using and developing RBL materials.

Table 1: *Examples of references to RBL in the other GDN Guides*

GDN Guides	Examples
Teaching and Learning	Development of RBL packages in consultation with an Educational Developer
Lecturing	Putting lectures on WWW; resource-based active learning exercises
Small-group Teaching	Library of tutorials; use of study skill packages in tutorial classes
Practical and Laboratory Work	Guides to laboratory techniques; health and safety guides; case studies of practicals using RBL
Fieldwork and Dissertations	Fieldwork booklets; dissertation guides
Teaching and Learning Geography with ICT	Computer assisted learning packages
Transferable Skills and Work-based Learning	Transferable skills study guides
Assessment	Assessment of student workbooks on techniques in geographical analysis
Curriculum Design	Virtual Geography Department; course guides

Many of the good teaching, learning and assessment practices identified in the other GDN Guides make use of RBL (Table 1) and this Guide makes cross-references to them. As there is a separate Guide on *Teaching and Learning Geography with Information and Communication Technologies* (ICT) (Shepherd, 1998), I give most attention to the use of non-IT resources.

Probably the best known example of an institution making extensive use of RBL is the Open University and an example of its use of RBL in geography is considered later (Box 24). However, as most geography departments, at least in the UK, have not attempted to apply this approach, I put the emphasis on the use of RBL to help students who attend a 'university campus' to engage in the majority of their learning and teaching, rather than those who are learning predominantly at a distance. The proportion of a student's time using materials varies from a low amount in a traditional course to a high amount in a distance learning programme, but it is increasing with greater use of RBL in campus-based courses (Figure 1). I would go along with Rowntree (1997, p.2), who suggests that "*any* shift from whatever is your present position towards greater dependence on materials (and less on face-to-face contact) may open up many new learning opportunities — while at the same time presenting a number of new challenges and potential problems."

Figure 1: The RBL continuum (based on Rowntree, 1997, p.2)

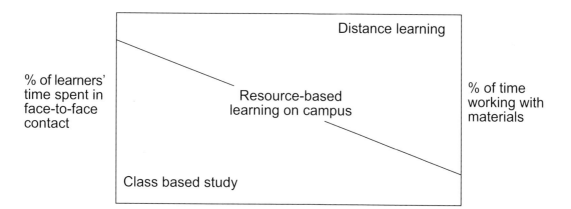

Activity 1

What *different uses of RBL* are you familiar with? Write below three different examples of the use of RBL in your Department (or elsewhere if you prefer).

1.

2.

3.

My main aim in preparing this Guide is to examine and illustrate how RBL may be used to enhance learning and teaching in geography. The structure of the Guide should help unpack the different forms of RBL and provide a framework by which you can evaluate the benefits and problems of developing and using RBL. I also attempt to give some general guidance on designing learning packages to promote student learning, ways of supporting students and staff using RBL, and strategies for introducing RBL into your courses and your department. I make frequent reference to examples of specific practices taken from degree courses, usually in geography, from a range of Higher Education Institutions (HEIs), mostly in the UK, but also some from North America and Australasia.

It was apparent in preparing this Guide that at present RBL packages are more extensively developed and used in the 'new' universities and colleges in the UK than in the 'old' universities. This may in part reflect that up until now the old universities have been under less pressure from increases in student numbers to change their learning and teaching methods and have the benefit of better resourced libraries. However, it is clear from the recommendations of the Dearing Committee (NCIHE, 1997) that RBL should form an important feature of learning and teaching in the future of all HEIs in the UK (Box 2).

Box 2:   RBL is central to the recommendations of the Dearing Committee on promoting student learning

Recommendation 8

We recommend that, with immediate effect, all institutions of higher education give high priority to developing and implementing learning and teaching strategies which focus on the promotion of students' learning.

8.13 …To manage the learning process for more diverse and greater numbers of students, teachers will have to consider the trade-off between the quantity and the quality of time spent with students. Planning for learning means that designing the forms of instruction which support learning becomes as important as preparing the content of the programmes.

Appendix 2 — New approaches to teaching — comparing cost structures of teaching methods

13. It is clear… that continued expansion will require a radical shift in the way university teaching is carried out. The implications of the 'Future' model are far-reaching. To achieve that kind of cost curve, while preserving staff-student contact, it was necessary to increase RBL to cover the majority of student learning time…

Reference:
    NCIHE (1997); see also Section 6.3

1.2   What are the Learning Objectives?

This Guide attempts to cover all the main issues concerned with the development and use of RBL in geography in higher education. If you work your way through all this Guide you should be able to each of the following (to help you identify which sections will be of most interest I have indicated the relevant section numbers in parentheses):

- describe the range of forms that RBL may take (Sections 2 and 5);

- identify the benefits and problems for you and your students of using RBL (Section 3);

- explain the main components of how students learn and apply them to the development and use of RBL (Section 4);

- evaluate the variety of ways in which RBL may be used to enhance learning in geography (this is the key objective and is the theme which runs right through the Guide; it is particularly emphasized in Sections 3, 5, 6 and 9);

- choose appropriate cost effective strategies for you and your department's use of RBL (Section 6);

- feel more confident about designing RBL packages and appropriate learning activities and assessments to promote learning (Section 7);

- assess the support that students and staff using and developing RBL need (Section 8);

- evaluate the relevance of the case studies of ways in which RBL is used in geography to you and your department (most of the boxed examples in Sections 3-8 deal with this, Section 9 presents five longer case studies).

Activity 2

How important are these *learning objectives* to you?

Re-read them and rank them from 1-8 in order of relevance to you.

1.3   Who should use this Guide, when and how?

Like all the Guides in the series this one has been written primarily for lecturers in geography and related disciplines working in HE and educational developers who work with staff in these disciplines (Healey, 1998a). However, most of the issues discussed are applicable to all HE lecturers and many of the same issues face staff in Further Education and Sixth Form Colleges. This particular Guide should also be of interest to academic support staff in academic libraries and learning centres. The main focus is on the situation in the UK, but most of the issues and practices are applicable to other countries, particularly North America and Australasia (Box 3).

Box 3:   **Readers may have a variety of reasons for looking at this Guide**

For example:

- You are an *experienced geography lecturer* who is in charge of revising a core introductory module, which next year is likely to recruit 200 students, a third more than the current version. You wish to examine the potential of RBL to improve student learning by introducing some active learning exercises in your lectures and to replace some lecture topics altogether.

- You have recently started your career as a *lecturer in environmental science* and wish to examine the benefits of using and developing RBL both for your students and yourself.

- You are the *geography course leader* developing a learning and teaching strategy for your department and you want to evaluate the potential of RBL to help you meet the challenge of providing a set of stimulating courses for the expansion in student numbers that you are attracting and the increasingly diversified backgrounds of the students starting your course.

- You have just been appointed *Chair of the new Department of Geography and Environmental Management*, a product of the latest institutional reorganization. Faced with overlap between existing modules and relatively poor recruitment to the Environmental Management course, you want to rationalize the number of modules on offer and you want to know whether or not it is worth investing the time and effort in converting some, or all, of the content of the revised modules to RBL.

- You are the *module tutor for GE201 Techniques in Geographical Analysis*, a compulsory module taken by 110 single honours geographers, 35 joint geography honours students and 25 students taking environmental studies. Student evaluations show that it is one of the least popular modules and only 40-50 per cent of students turn up to the weekly lecture after the first two weeks. You are considering converting your lectures into a self-instructional guide linked directly to the weekly practical and replacing the weekly assessed practical exercises by computer marked weekly tests. You want some advice on developing an effective study guide.

- You are the *University Educational Developer* who, in line with your institution's new learning and teaching strategy is encouraging RBL. You are organizing a series of staff development workshops on independent learning for related disciplines. You wish to examine current practices in geography.

- You have been given the job by your Head of Department of *developing a learning package on survey techniques and data sources in human geography* aimed at students undertaking dissertations in this area. With the increased student-staff ratios staff are finding it increasingly time consuming to give adequate individual advice to students on these topics. You want to learn from the experience of other geographers who have developed learning packages.

To reflect the varied interest of readers I have tried to design the Guide so that it can be dipped into when and where appropriate. It is structured in two main parts. The first, and longest part from Sections 2 to 8, focuses on the issues involved in using and developing RBL. The issues are illustrated with numerous examples (shown in boxes) of over 50 practices in a wide range of mainly geography departments. Section 9 contains a collection of more detailed case studies which illustrate a range of ways in which RBL is currently used in geography. You may prefer to start with the case studies and look at the examples in the boxes and then come back to examine pertinent issues which are raised by them and reflect your needs. Section 5 is particularly valuable in identifying the range of ways in which RBL is used in geography to enhance learning and you may wish to start with this section. A list of the boxed examples is given at the beginning of the Guide.

Twenty-five *activities* have also been included in the text. These are designed to assist your reflection on the relevance of the material to your situation. Please spend a few moments jotting down your thoughts on these activities, because it should help you take ownership of the ideas and decide on their relevance to you. However, I recognize that people have different learning styles and that different activities will appeal to different people. So please have a go where they seem useful to you, and miss them out where they are not relevant for your purpose.

I have designed a workshop to go with the Guide (Box 52), details of which may be obtained from the GDN Project Officer (see front of Guide). The Guide has, however, been written to be used independently of the workshop.

Activity 3

What is your *motivation* for reading this Guide? Write below up to three reasons why you are looking at this Guide.

1.

2.

3.

Look at the *Contents* list at the beginning of this Guide and decide which sections are relevant for your purposes and in what order you wish to look at them.

2 But what *is* resource-based learning (RBL)?

Some people revel in discussions of definitions and terminology, others prefer to move straight on to the main issues. If you would like to clarify the different uses of the term resource-based learning and associated terms read on; if, on the other hand, such debates are not to your liking skip to the last paragraph of this section which gives the working definition of RBL used in this Guide.

RBL is difficult to define, as was noted in the last section. At its broadest it can include any resource which students use in their learning, including lecturers and other students (for example, see definition 2 in the Activity 4 Box). Defined in this way it can include most forms of learning and teaching. This is unhelpful for our purposes because we are producing a series of Guides on different aspects of learning and teaching. To distinguish this Guide from most of the other Guides in the series, resources are interpreted to mean print-based, audio-visual and computer-based materials. However, most attention is paid to the print-based resources, because there is a separate Guide on *Teaching and Learning Geography with Information and Communication Technologies* (Shepherd, 1998) and there is, as yet, relatively little use by students of audio and visual resources in geography in HE, independently of print or computer-based media (see Gold *et al.*, 1996 (Box 20); Jenkins & Youngs, 1983; Whitelegg, 1982; and Section 9.1 for some examples of the use of audio and visual resources in geography).

Activity 4

Definitions of RBL in part reflect the perspective of the authors and the audience they are addressing. Below are *four definitions of RBL*, two are by librarians and two by educational developers. Which do you think is which?

- "Resource-based learning actively involves students in the complex process of recognizing the need for information, identifying and finding the relevant information, evaluating it, organizing it, and using it effectively to address problems."

- "Most learning is… resource-based learning. Learning resources take many forms, and include human resources (tutors, mentors, fellow-students), and information-type resources (books, databases, on-line databanks, learning packages, lecture notes, handouts, manuals, and so on)."

- "Resource-based learning goes beyond classroom lectures and textbooks to draw on a wide variety of information sources and formats, both within and outside libraries."

- "RBL is difficult to define. The traditional academic library is the major learning resource in most institutions, and, in a sense, conventional courses, where students attend lectures and then use the library to support their studying, involve a form of RBL. …It is also possible to describe tutors or even students as learning resources. However the definition we are using is narrower: for our purposes, RBL is the use of mainly printed materials, written, collated, or signposted by tutors, as a substitute for some aspects of teaching and library use."

The sources of the quotes are given in Appendix I

There are many other terms in the literature, such as distance learning, open learning, flexible learning, problem-based learning, and independent learning, which are used by some authors interchangeably with RBL. Other writers make distinctions between the terms (Box 4).

Activity 5

Given that the definition of all these terms is problematic you may like to note down what you understand by them before looking at the box below:

Distance learning:

Open learning:

Flexible learning:

Problem-based learning:

Independent learning:

Resource-based learning:

Box 4:   There is a confusing range of RBL related terms in common usage

Distance learning describes "a course offered to learners who study largely by themselves — at home or at work or elsewhere — with little or no opportunity to meet other learners or tutors" (Lewis and Freeman, 1994, p.123). "However, all students, whether college based or not, using self-instructional materials are to some extent distance learners" (Hodgson, 1993, pp.40-42).

Open learning is a broader term and includes distance learning. It describes "an approach to learning provision which emphasizes learner choice as a major determinant of course design. Open learning is characterized by offering the learner choice over aspects such as time, pace and place of study, but no single collection of features defines or is common to all open learning programmes" (Lewis and Freeman, 1994, p.124).

(cont.)

Flexible learning is an even broader term, and "includes all forms of open and distance learning, but also includes other learning situations which at first sight appear more traditional. For example, 200 people in a lecture theatre can be 'learning flexibly', if (for example) they are spending five minutes where everyone is working through a handout and trying to answer some questions. In other words, flexible learning involves people taking some control regarding how they learn" (Race, 1993, p.92).

Problem-based learning "provides problems to be tackled, through which learning takes place. For the learner the purpose of the activity is clear even if the learning outcomes are not yet obvious. People seem to have a natural propensity to tackle problems and find them inherently engaging. In some forms of problem-based learning there is a detailed analysis of the knowledge base the problem draws upon, to guide learners, and most problem-based learning also involves groups who help to give directions to learning" (Gibbs, 1992a, p.51). It is most commonly used in professional education, such as medicine, law, social work, engineering and management.

Independent learning " describes a wide range of practices. It has become a rallying cry for those who believe that students need, or can cope with, much less support from teachers than they often receive, and that such independence is beneficial to students. ...Independent learning nearly always involves extensive independent use of the library and other information sources rather than formal teaching. Lecturers' time is concerned more with identifying clear learning goals, providing support and feedback during learning, and assisting in the collation, presentation and assessment of learning outcomes than with conventional teaching" (Gibbs, 1992b, pp.41-42).

Resource-based learning occurs "when individuals or small groups of people learn from things such as self-instructional materials, textbooks and apparatus or exhibits of various kinds. Open and distance learning may use any of these, and are thus, by definition, forms of resource-based learning. Not all resource-based learning systems are suitable for distance learning, though, because many of them require the presence of a tutor to carry out such tasks as organizing small groups, doing oral work or practical demonstrations, providing remedial tuition, supervising stage tests, etc. The resources for such learning systems are often kept in one location such as a library or resource centre and the learners come to that location to use the material" (Hodgson, 1993, p.108).

Although there are differences in the way in which the term RBL and other related phrases are used, all emphasize that this is a form of learning which is student-centred rather than teacher-centred. In other words, resource-based *learning* is characterized by the use of resources *by students*. Where it is *the teacher* who is using the resource, for example, showing a video or running a computer simulation in a lecture, this is normally resource-based *teaching* (Box 5).

> **Box 5:** **Resource-based teaching and resource-based learning are not the same thing**
>
> "In resource-based teaching, the teacher is using resources to broaden his or her instructional base. In addition to the lecture and textbook, the teacher may make use of other print resources such as newspapers or magazines, and audiovisual resources such as films or videos, as well as other human resources, such as guest speakers. Certainly one might argue that the learning environment is enhanced through this approach. But the teacher is still at the center (sic) of that environment. The focus is on what the teacher is doing with those resources to facilitate his or her teaching. And such media-driven lessons are generally still geared to passive student absorption rather than inquiry.
>
> In resource-based learning, students use resources to broaden their learning base. They may access all the same kinds of resources referred to in our resource-based teaching example, including the teacher and the textbook. But the students are the center (sic) of the learning environment. The focus is on what the students are doing with those resources to facilitate their learning" (Haycock, 1991, p.16).

RBL is a more commonly used term in Britain than it is in North America, where the term 'active learning' is more frequently used. However, active learning is a broader term, which reflects a general student-centred approach to learning. RBL shares this approach, but with much active learning the *only* resources used are other students and the teacher. Whereas with RBL, as used here, the focus is on material-based resources (see Section 7.1.3).

One of the reasons it is difficult to define RBL and to identify its characteristics is that the way in which it is used, for what purpose, in what form, and by whom varies widely. Figure 2 (overleaf) illustrates some of the spectra along which different RBL packages lie.

Activity 6

- Take two real or imaginary examples of learning packages (e.g. a project learning exercise, a dissertation guide, a set of word-processed lecture notes, a reader, or a video demonstrating the use of survey equipment) and using Figure 2 mark the relative position of each package along each of the RBL spectra.

- Think about the circumstances when other learning packages might be designed to fall in other positions along the spectra.

To recap, for the purpose of this Guide, RBL is taken to mean the use by students of print and electronic-based learning resources (but primarily the former) in- and out-of-class as part of learning exercises and packages devised by lecturers in HE.

Figure 2:   *Resource-based learning: the spectra (Source: Roberts, 1995)*

Conventional delivery	Partial substitution	Distance learning
Direction and pacing with tutor	Some deadlines	Direction and pacing with student
Passive learning		Active learning
Frequent tutor contact	Limited tutor contact	No tutor contact
Two-way dialogue	One-way dialogue	No dialogue
Classroom base	College/Learning Centre base	Home base
Institution holds material		Student holds material
Academics deliver		Administrators deliver
Single medium	Multimedia	Multimedia with TV
Low technology		High technology
Instruction		Education
Introductory/ factual material	Discursive/ debatable	Abstract/ philosophical
Subject orientated aims		Personal development aims

3 How can RBL help you and your students?

When suitably designed and supported, RBL has the potential to help lecturers meet some of the challenges facing HE, and provide students using the materials with many advantages. There are, however, some potential problems with developing and using RBL materials which need to be overcome or minimized.

3.1 How can RBL help meet the challenges facing geography lecturers in higher education?

The main issues facing geography lecturers in the UK — getting more for less; enhancing quality and preparing for internal and external quality audit; maintaining standards; dealing with student diversity; the challenges of modularization and building in progression; demands of the market place; and bringing it altogether — are reviewed in another Guide in this series (Gardiner & D'Andrea, 1998). The effects that the declining unit of resource and the increased emphasis on research has had on the quality of learning and teaching in geography are reviewed in Healey (1997). Similar challenges face lecturers in many other countries. RBL has an important role to play as one part of a learning and teaching strategy for responding to many of these issues (Gold, *et al.*, 1991).

3.1.1 Large classes are not working well

The problems caused by larger classes in HE have been extensively discussed in recent years (Cryer & Elton, 1992; Gibbs & Jenkins, 1992a; OCSD, various). These include:

- fewer opportunities for discussion;
- a reduction in feedback;
- more students who are ill-prepared;
- increased reliance on teacher input;
- a greater emphasis on 'surface' learning;
- increased absenteeism;
- a negative impact on grades;
- greater use of inappropriate and poorly equipped rooms.

There are a range of strategies for responding to these problems, but the observation of Her Majesty's Inspectors (HMIs), who used systematically to observe and comment on classes in the new universities and colleges in England, still rings true today: "In many institutions… tutors continue to organize teaching and learning in ways which are no longer compatible with the numbers of students involved" (DES, 1991, p.21; quoted by Gibbs & Jenkins, 1992b).

Race (1993a, p.113) argues that: "If we focus on learning rather than on teaching, we are on our way towards helping greater numbers of students benefit from higher education. In short, if we look after the learning, the teaching will look after itself — whatever the staff-student ratios." RBL can contribute to reducing some of the problems associated with large classes (Box 6) by, for example:

- giving greater access to learning materials;

- increasing the amount of active learning;

- providing immediate feedback;

- encouraging 'deeper' learning;

- reducing pressure on classrooms.

> **Box 6:   RBL can alleviate some of the problems associated with large geography classes**
>
> At the University of Texas at Austin, the Urban Geography course (315) has been delivered in RBL form for the last 18 years. It is designed as a self-paced package and meets each of the five bullet points, listed above, for reducing problems associated with large classes. The students study eight core units and two out of six option units. Each unit requires the students to complete a reading assignment and pass a test. There is a study guide to the units and readings, which consist of a core text, an issue of a journal and four articles. The only occasion the class meets all together is in the first week, at which time the details of the course are discussed and advice is given on how to make a success of it. Most of the tests are a mixture of 40 questions worth point each, or short essays, worth between three and seven points each. Several testing times are available during the week and the students may take the tests when they are ready. To encourage the students to pace themselves all of them must complete at least two tests at an early stage during the course, or suffer a points penalty. Bonus points are given for completion of the tests a month before the last testing session. Approximately 100 students take the course each year. As the method of learning is new to the students, many do not like it to begin with, but as they get used to it they appreciate the flexibility. Students with queries or difficulties can meet with the Course Leader or the Teaching Assistant.
>
> *Further information*:
>   Chris Davies, Department of Geography, University of Texas at Austin; email: horace@mail.utexas.edu

3.1.2   Students are more diverse

As HE has expanded, the background of students studying geography courses has diversified. Students in England, Wales and Northern Ireland who are not 18-20 year old with an A-level in geography are no longer an exception and in many institutions they form a sizable proportion of the entry profile. An increasing proportion of geography students are mature, many of whom have come from Access courses. Part-time students, although by no means as important as in many other subjects, have also increased. Perhaps more significant though is the way in which the characteristics of the full-time traditional student body is changing with the rapid growth in the proportion of students who are working during term-time (Gardiner & D'Andrea, 1998). Under these circumstances, courses in which every student is presented with the same information at the same rate, primarily through lectures given once a year in a fixed time slot, are unlikely to meet the needs of all the students or to work successfully.

RBL packages can help provide for the needs of the more diverse group of students taking geography courses by:

- providing flexibility in the time (and sometimes place) at which study occurs;

- enabling students to study at a speed which suits their background knowledge, the relative difficulty of the material and their favoured way of learning (Box 7);

- filling gaps in the knowledge they bring to a course.

Box 7:   RBL is an important way of providing for the varied needs of the more diversified body of students studying geography

At Kingston University the RBL package on Social Surveys and Geographical Investigation, which takes up one week's work, is particularly liked by part-time students registered for the module, because it means that there is one less evening on which they need to physically attend the campus.

Further information:
  Bob Gant, School of Geography, Kingston University
  email: r.gant@kingston.ac.uk

3.1.3   Increased student-staff ratios (SSRs) are reducing the amount of tutorial and supervisory support which can be afforded

Between 1986 and 1991, SSRs in British University geography departments rose by 44 per cent, from 12.1:1 to 17.4:1 (Jenkins & Smith, 1993). They have continued to rise since then. Much of the individual, and some of the small group, tuition and supervision has had to be rationed or is no longer viable. This is particularly true for 'informal' out-of-class contact. The most common complaint from students after 'the book is out' is 'my tutor is always out' (Gibbs *et al.*, 1994a). Some staff and some departments have instituted 'office hours', a practice already common in North America. Specified, limited times when they are available for student consultation are posted on their office doors. As students can no longer get the advice they want *when* they want it, many departments are anticipating commonly asked queries and are providing an increased number of written guides and handouts. These are probably the most common form of RBL in use in HE.

Box 8: **A variety of guides are commonly provided for geography students**

Examples of guides provided to geography students at Cheltenham and Gloucester College of Higher Education include:

- *A Student's Guide to the Department*
- *Field Guide for Geography, Human Geography, Physical Geography, Geology and Natural Resource Management: Level 1; Level II; Level III* (includes aims, learning objectives, weekly timetable, form of assessment and deadlines, and list of key texts for every module)
- *A Staff and Student Guide to Academic Counselling*
- *A User's Guide to the Laboratories*
- *Department Health and Safety Regulations*
- *Acknowledging Sources in Assignments*
- *Dissertations in Human Geography Field*
- *Dissertations in Physical Geography Field*
- *Dissertations: A guide to important deadlines and preliminary procedures*
- *Dissertations: Guidance notes on dissertation format*
- *Writing a Field Notebook*
- *Independent Study Module: Introduction and guide to deadlines and procedures*

Further information:
Margaret Harrison, School of Environment, Cheltenham and Gloucester College of Higher Education; email: mharrison@chelt.ac.uk

3.1.4   Need for staff to find more time for other activities

Academic staff in HE are under increasing pressures to perform as effective researchers, generators of income, and administrators, as well as being capable teachers, facilitators of learning and student counsellors. Providing written Guides and operating 'office hours' are two ways that you can manage the 'out-of-class' time you spend with students. RBL may also be used to substitute for 'in-class' time. There is evidence that students "often learn more efficiently from reading than from listening" (McKeachie, 1994 p.129). Although you may need to invest time 'up-front' in preparing a learning package, using or modifying existing resources can reduce this investment significantly. The savings in time come when you use the package. The more frequently the package is used, instead of you being present, the greater the savings. Spreading the pressure on staff resources more evenly over the year is another advantage of using RBL. This was one of the factors which led Plymouth University to use RBL for a new final year core module (Jones, 1986). It is important that the costs and benefits of developing and using RBL are calculated carefully (see Section 6.4).

The time you save can then be used to enhance teaching where direct contact with students is most needed, perhaps in small group discussions, or you may use it for non-teaching activities (Boxes 9, 25 and 26; Case Study 9.1). Having a stand-alone, self-learning package available which can be slotted into your course when you cannot be present, perhaps because of another commitment or illness, provides you with greater flexibility. It avoids trying to get a colleague to cover the session at the last minute, having to cancel the class, or not being able to take up the opportunity offered. If that invitation to participate in a seminar in Hawaii does not appear you can, of course, still use the package at an appropriate time during the course and use the time saved to do something else.

Box 9: Using a learning package can provide staff time for other activities

A Social Research reader has been produced at the University of Hertfordshire for environmental studies students to replace much of the repetitive supervisory work. Students are expected to consult the reader before seeking help from their dissertation tutor. This gives more supervisory time for discussion of data interpretation and analysis.

Further information:
Jennifer Blumhof and Debbie Pearlman, Division of Environmental Sciences, University of Hertfordshire; email: J.R.Blumhof@herts.ac.uk

Reference:
Exley & Gibbs (1994)

3.1.5 Need to provide consistency and quality

Many practical, seminar and tutorial classes are repeated several times for large classes, particularly in introductory courses, and are run by more than one member of staff, often including postgraduate teaching assistants. Under such circumstances both staff and students can gain from having a consistent set of activities. Moreover, where the composition of the teaching team can change from year to year, such guides can provide a helpful induction to the course and make the changeover much easier and economical (Box 10). The same argument of ensuring that the quality of learning is maintained and controlled applies to developing learning packages for use in franchised colleges.

Box 10: A set of prepared activities can give students a consistent experience and save staff considerable time where the same class is repeated several times

Geographers at Lancaster University have established a library of tutorials for use by staff and postgraduate tutors. Each tutorial is complete with rationale, aims, handouts, details on how to run it, reading material and other resources, and briefing notes for tutors less familiar with the subject matter. The tutorials can be used immediately or in a modified form to suit individual tutors' expertise. A side benefit is that there has been an informal convergence of the tutorial programmes used by tutors without the difficulties of devising and imposing a single rigid syllabus.

Reference:
Clark & Wareham (1998, p.21)

Quality is often enhanced by developing learning packages. Most tutors will take more trouble over the content and presentation of materials which are more open to 'public' scrutiny, particularly from colleagues. Hence the quality of the materials used in RBL are usually higher than the notes that tutors use in their lectures, practicals and tutorials to present the same ideas.

Activity 7

Review the issues facing HE (Section 3.1) and rank them in order of importance to your situation. Are there other issues which you face which RBL may help you with?

- Large classes are not working well
- Students are more diverse
- Increased student-staff ratios (SSRs) are reducing the amount of tutorial and supervisory support which can be afforded
- Need for staff to find more time for other academic activities
- Need to provide consistency and quality

3.2   What are the benefits for students of using RBL?

The issues facing HE, identified in Section 3.1, were dealt with from a staff viewpoint. Many also apply to students. In this section the advantages of using RBL are considered explicitly from the point of view of the students using the materials.

3.2.1   Promotes active learning

Educational research and practice shows that a key characteristic of good teaching is the provision of plenty of opportunities for active learning, through getting students to think, do and reflect upon what they are doing (Gamson & Chickering, 1987; Ramsden, 1992). Good RBL packages provide many such opportunities (Boxes 11, 42 and 45). When the materials focus simply on content or advice it is important that an active element is built into exercises undertaken in or out-of-class, if they are to provide effective learning opportunities (see also Sections 6.5.4, 7.1.3 and 7.2).

Box 11:   Active learning is at the centre of effective RBL geography packages

Clark University in association with the Association of American Geographers (AAG) have developed ten active learning modules on *The Human Dimensions of Global Change*. Designed for use in any introductory course that deals with human-environment relationships, the modules actively engage students in problem solving, challenge them to think critically, invite them to participate in the process of scientific enquiry, and involve them in cooperative learning.

Each module consists of several units, each of which focuses on an aspect of the module's theme. The core of each unit is a variety of student activities that have been designed to be challenging, but not baffling. The activities vary in type, in the time they require, in skill level assumed, and in the skills developed. They involve critical reading; data collection, assessment, interpretation, and analysis; map reading and interpretation; field trips; interviewing; role playing; and writing for particular audiences. Many activities link the students' own lives with processes of local, regional, and global change. Each unit comes with some background reading to introduce the topic.

The modules provide instructors with a broad array of specific ideas for involving students actively and collaboratively in learning about nature-society relationships; instructors can choose the activities that best suit their class size, students' abilities, and instructional settings. Some activities can be completed in one class session whereas others are out-of-class projects and may take longer. An instructor who uses an entire module will have material for roughly two weeks of classes.

Reference:
   Hands-on! Project Overview and Module Summaries (see also Section 10 — Guide to references and resources)

3.2.2   Involves students in classes

Where RBL is integrated into a class, it can provide an opportunity for students to learn individually and/or through cooperation with other students. The resource materials in this case are used to stimulate a learning activity, such as the interpretation of some data or a discussion of contrasting viewpoints (Box 12). Several examples of the use of such RBL exercises are given in the GDN Guides on *Lecturing in Geography* (Agnew & Elton, 1998) and *Small-group Teaching in Geography* (Clark & Wareham, 1998).

Box 12:   'One of the best ways of learning is to teach someone else'

Teaching each other is an effective way of learning which can be used to good effect in a lecture or as part of a tutorial class. As an alternative to the students passively receiving a summary from the lecturer the students act as tutors to each other. In this exercise students, usually working in pairs or threes, are each given a different piece of written material, for example, a different case study or extract from an article (two or three pages maximum), which they have to study. Each student then has a minute or two to 'teach' the other member(s) of their group the key points that they have abstracted. The exercise can continue by asking the group, for example, to draw out the similarities and differences between the different pieces of information, or to synthesize their findings and apply them in another context.

Further information:
   Mick Healey, Geography and Environmental Management Research Unit, Cheltenham and Gloucester College of Higher Education
   email: mhealey@chelt.ac.uk

Reference:
   GDN WWW Resource Database
   http://www.chelt.ac.uk/gdn/abstracts/a68.htm

3.2.3   Provides feedback

Where RBL is integrated into a class, the tutor can provide immediate feedback by, for example, summarizing the main points or commenting on the issues raised by say, two students reporting back on their findings. In learning packages studied independently it is important that a mechanism for providing feedback is given (Box 13). This may be, for example, through a later tutorial class, where the issues are going to be discussed, or by providing some commentary elsewhere in the package on the issues raised (see also Section 7.2.2).

Box 13:   Self-assessment questions (SAQs) are an essential component of independent learning packages

In the GIS Professional Training Programme at Kingston University, which is delivered as a distance learning package, SAQs are used throughout the modules. They are designed to help the students think through some of the topics discussed and allow them to include any of their own GIS or related experiences. A suggested time limit is given for each question. Feedback is provided by giving some sample answers to the SAQs at the end of each unit.

Example: Module 1 Unit 1 *SAQ 2: The use of databases*

Can you give examples of any databases that you know of in your working environment?

In addition do any of these databases contain geographically referenced information?

Sample answer: Based on Kingston University Databases:

MIS (Student records); (Health Centre); Customized Oracel (Personnel); Quattro Pro (Payroll); ITALIS (Library)

Geographical referencing: Yes, all of them give address and post code information.

Further information:
Teresa Connolly, GIS Professional Training Programme Manager, Kingston University; email: t.connolly@kingston.ac.uk

---

3.2.4   Increases access to resources

As student numbers have increased, library expenditure budgets have not kept pace. Whereas the average spend per student on books in university libraries in the UK was four books a decade ago, now it is one (Brown & Smith, 1996a). Although libraries have taken various measures to alleviate some of the pressure on the most popular books, obtaining access to key books and journals is increasingly difficult for many students, particularly in large classes. Several institutions have developed readers to overcome this problem (Boxes 14, 49, 50 and 53). RBL packages which include a copy of key articles and chapters and other resources, such as an audio or video tape is another way in which access to learning resources can be improved. Care needs to be taken that the relevant copyright permissions are obtained (Appendix II).

> **Box 14: Module readers are an important way of improving access to library resources for your students**
>
> In the late 1980s and early 1990s Middlesex University developed five readers to support modules on Global Environments, Culture and Development, Heritage Studies, Environmental Modification, and Environment and Development. Each one consisted of a bound volume of printed materials which was given a degree of coherence by the addition of linking text. The readers were sold to students at cost price.* In an evaluation, involving over 70 students across the three years of the geography course, Shepherd & Bleasdale (1993, p.112) found that: "Overall, students most frequently praised the readers for the time they saved in tracking down materials in congested libraries, for the convenience of having essential reading within one pair of covers and for providing a comprehensive, general coverage of module themes."
>
> *Reference*:
>   Shepherd & Bleasdale (1993)

3.2.5   Provides greater flexibility

RBL packages designed for independent learning provide students with flexibility as to when, and usually where, they use the package. They also enable students to go at their own pace, skimming through material they find easy, or are already familiar with, and spending longer on the sections which are more difficult, or new to them. The ability to go at their own speed is particularly important for students for whom English is not their first language, who often find that material presented orally is delivered at too high a speed for them fully to comprehend.

> **Box 15: The same RBL materials may be used by geography students in a variety of ways**
>
> In a course on Water Supply at Bradford University audiotapes and session notes have been used to replace lectures by seminars. "The students study the tapes at their own pace. Approximately half of them work totally independently and the others work together in informal self-help groups. Groups of three or four students can often be seen listening to the tapes together in the learning resource room" (Exley & Gibbs, 1994, p.55).
>
> *Further information*:
>   Bob Matthew, Department of Civil Engineering, Bradford University; email: r.g.s.matthew@bradford.ac.uk
>
> *Reference*:
>   Exley & Gibbs (1994); plus updated case study in Section 9.1

---

* *Changes to the interpretation of the copyright law in England led Middlesex University to decide it was too expensive and time consuming to obtain permission to produce readers like these, which consist of several different articles along with interpretative comment bound together in a collection which are then sold to individual students, and the Department has ceased their production. Other institutions continue to produce readers under their agreements with the Copyright Licensing Agency (see e.g. Boxes 49 and 53). See also Appendix II.*

3.2.6   Develops independence and life-long learning skills

Using RBL materials gives students experience of working independently of a tutor, whether on their own or in a group. Managing their own learning is a key skill. Indeed, the learning that most people do when they have left college or university is resource-based and student-centred; therefore assisting your students to learn in this way helps to prepare them for life-long learning (Box 16).

Box 16:   Learning to search effectively and efficiently for information is one of the most important life-long learning skills

At the University of Sussex, as part of the first year Geographical Methods course, they use the *GeographyCal* module on 'Making Sense of Information' to develop effective and critical information search strategies among the students, emphasizing particularly the potential and pitfalls of using the WWW. A one-hour lecture was given designed to outline the geographical relevance of using the Internet; this was backed up by a three hour workshop and a seminar discussing issues arising from the exercise. Some students found the workshop session too long despite the availability of generous coffee breaks. Other students refused to take a break because they became very attached to their newly acquired WWW browsing skills (though not always in search of geographical material!).

When the course ran for the second time, in the Summer Term 1998, the formal lectures and seminars were discontinued. Instead a two hour practical was used in which the topic and material were introduced, followed by a three hour practical in which the students embarked upon the CAL and targeted Internet material to answer a series of graded questions. This structure gave the students more time to explore the material. Small seminar-like discussions with groups of students occurred spontaneously as needs arose and interests were expressed.

Further information:
  Tom Browne, Computing Services, and Don Funnell, Department of Geography, University of Sussex; email: t.j.browne@sussex.ac.uk

Reference:
  Funnell & Browne (1997)

Activity 8

- Look back at the above *benefits for students* of using RBL (Section 3.2) and rank them in order of importance for the students whom you teach; if possible ask a group of your students who have used RBL packages to do the same exercise and compare their rankings with yours.

- Are there additional benefits which your students gain from using RBL materials which are not discussed here?

3.3   What are the potential problems with developing and using RBL materials and how may they be reduced or overcome?

Despite the potential advantages for staff and students of using RBL, problems may arise during the development and use of RBL materials which need to be overcome or minimized. This section is largely based on Gibbs *et al.* (1994a, pp.8-10).

3.3.1   It is not enough to write down the content

To obtain the best from RBL materials considerable attention needs to be paid to the process as well as the content. Appropriate learning activities, assessment exercises, support and feedback need to be built into the materials (see Section 7).

3.3.2   There is still need for contact with and between students

Although managers may be attracted by potential savings from replacing expensive staff-student contact with RBL materials, relatively few of the students we teach are sufficiently self-motivated, self-disciplined and bright to work entirely independently. The Open University abandoned its experiments with 18-21 year old students because their completion rates were so low. Contact between staff and students is still vital for four main reasons:

- motivation;
- pacing;
- sorting out problems;
- understanding.

though the time involved may be considerably less in comparison to traditional teaching methods. Open University courses usually involve more coursework and tutor feedback than face to face courses do as a way to pace students through (see also Section 6.2). Contact between teacher and learner is particularly important for the kind of learning conversation in which teachers adapt to the learner's understanding and misunderstanding of the material (Atherton, 1998; Laurillard, 1993).

3.3.3   RBL may be inflexible

Many RBL courses operate on fixed timetables with a tight weekly schedule. Although this can help students monitor their progress, it takes away some of the flexibility for students to study at times of their own choosing to fit in with their other commitments. There is also a danger of over-specifying material in learning packages so that students rely entirely on the package provided, do not purchase any other books, and have little incentive to go to the library to look for additional material. Flexibility has to be designed into packages, it does not come automatically.

3.3.4   Students may not know how to learn from resources

Most students have developed skills to cope with taught courses and think this is the only valid way to learn. We need to assure that our students develop the skills to learn effectively from resources, otherwise courses using RBL may run into problems (for example, Baxter, 1990a; see also Section 8.1.2).

3.3.5   Students may dislike learning from resources

Increasingly, as students have to pay for the higher education they receive and work their way through university, they have a greater sense of 'value for money'. This can lead them to be critical of RBL, although their reservations may be based on unsophisticated expectations about the nature of learning in higher education, rather than valid criticisms of the method itself. RBL courses are often better designed and more supportive than conventional courses, but it is essential that the reasons for using this approach are explained to our students and their anxieties are allayed. If RBL is a significant element of the course in the first year then students may accept the use of RBL units in later years more easily (Jones, 1986).

3.3.6   There may be a lack of institutional support

Teachers, accommodation and learning resources are used in new ways with RBL, and many features of institutional infrastructure can block developments in RBL. Institutional support is needed to help in the *production* of RBL materials (for example, from copyright librarians, computer-software designers, and desk-top publishing designers) and in the *use* of RBL (for example, provision of rooms for individual/group work, and good access to information search and retrieval facilities) (see Section 8.2).

3.3.7   The developmental time may be excessive

The time taken to develop RBL packages varies widely, according to the type of materials and the media being used (see Section 6.3). Using, or adapting, RBL materials produced by someone else can significantly reduce the developmental time (see Section 6.5).

3.3.8   RBL may not improve the quality of student learning

RBL does not automatically lead to an improvement in the quality of learning. It may be just as likely as other methods to produce disinterested students who underperform. "Whether the potential benefits are reaped or quality collapses depends on careful and thoughtful design, sufficient planning and preparation time, adequate resourcing, thorough implementation and a continuing cycle of evaluation and development" (Gibbs *et al.*, 1994a, p.10).

Activity 9

- Think about an example of how you might use RBL in your teaching.

- What problems do you think you may face in developing and using the materials?

- How can you minimize or overcome them?

4 How effective is RBL in enabling learning to take place?

A key issue, which needs to be considered before investing in the development of RBL materials, is how effective they are in helping students to learn. Not surprisingly there is not a definitive answer to this question. There are good and bad RBL materials, just as there are good and bad tutorials and lectures. However, consideration of how students learn, how RBL may be used to promote learning, and the methods of measuring effectiveness, should help you design effective learning materials.

4.1   How do students learn?

There are many theories of how students learn. One of the most widely adopted models, which is referred to in several of the other GDN Guides (e.g. Shepherd, 1998), is Kolb's Experiential Learning Cycle (Kolb, 1984). This provides a framework of the learning process — concrete experience, reflective observation, abstract conceptualization, and active experimentation.

Activity 10

Try answering the following questions and then look at the feedback in Appendix I.

1.   Think of something you're good at — something you know you do well. Write down a few words explaining how you became good at it.

2.   Think of something about yourself that you feel good about — a personal quality or attribute, something that 'gives you a bit of a glow'. Write down a few words explaining why you feel good about it. In other words, upon what evidence do you base your positive feeling?

3.   Think of something that you don't do well — for example, an unsuccessful learning experience. Write down a few words describing the causes of this unsuccessful learning experience.

Reference:
  Race (1993a)

Race (1993a) has proposed an alternative model using more everyday language (Table 2). Although the 'wanting, doing, feedback, digesting' model of learning may be thought of as cyclical (Figure 3), Race (1993a, p.14) prefers to emphasize the overlapping nature of the processes. "The 'wanting' stage needs to pervade throughout, so that 'doing' is wanted, 'feedback' is positively sought, opportunities for 'digesting' are seized, and so on."

Table 2:   *How people learn (Source: Race, 1993a, p.13)*

Wanting	motivation
Doing	practice
	trial and error
Feedback	other people's reactions
	seeing the results
Digesting	making sense of it
	gaining ownership

Figure 3:   *The wanting, doing, feedback, digesting model of how people learn (Source: Based on Race, 1993a, p.13)*

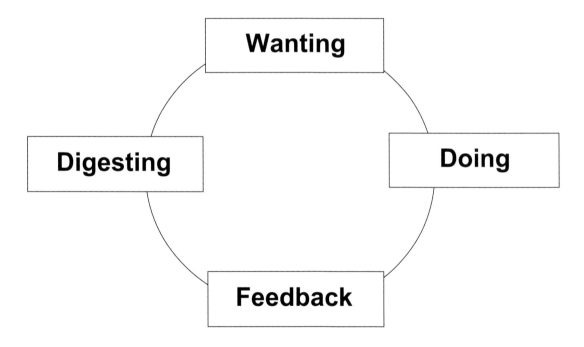

In a later article Race (1996) applied the 'wanting, doing, feedback, digesting' model to designing effective RBL. He shows how a variety of learning resources — handouts, computer conferencing and email, and videocassette recordings — can be used effectively by paying attention to the four factors. He also comments on the assessment considerations associated with each of the factors. The template for handouts is reproduced overleaf (Table 3).

Table 3: *How students learn and can be assessed using RBL: Handouts (Source: Race, 1996, p.34)*

Handout materials

Ways in which they can promote...	*Assessment considerations*
Wanting to learn	*Nature of learning outcomes*
Saving tedious note taking.	Establishing principal areas that students should learn.
Presenting 'digests' of important information.	Helping students to structure their learning of subject matter.
Including information about learning targets and objectives.	Defining details of the syllabus and its assessment.
Learning by doing	*'Doing' that can be measured*
Can include activities and exercises for students to do during lectures and in their own time.	Students' completion of exercises included in the handouts.
Can refer students out to textbooks and other learning materials.	Students' summaries made after studying handouts.
Can suggest that students make summary notes during or after lectures, or after further reading.	Students' answers to set questions involving them doing further reading or research.
Can form a basis for discussing ideas with fellow viewers.	Levels of participation in discussions of things covered in the handouts.
Can be a basis for participating in group debates.	The extent to which students apply critical thinking to the material in further tasks and exercises set after they have followed up the handouts.
Can be used by students for working out lists of questions of 'matters arising' from the handout.	
Learning through feedback	*Assessment and feedback*
Handouts can include self-assessment feedback responses to exercises and questions.	Tutors giving feedback about whether students have identified the most important features from the handouts.
Handouts can include 'expert witness' feedback from tutors on important questions tried by students.	Giving live feedback on students' answers to questions arising from the handouts.
Making sense of what has been learned	*Suggestions about assessment criteria*
Handouts can include reflective reviews of different interpretations and approaches to a topic.	Make sure that students know what they are expected to do with the activities and exercises in handouts.
Handouts help students to work out what are the most important factors on a topic.	Make the criteria explicit so students know what they are expected to get
Handouts can help students by including annotated bibliographies and reviews.	Where possible, negotiate the criteria with students, so that they feel a sense of ownership of the assessment agenda.

Activity 11

Select a learning resource (for example, audio tapes, World-Wide Web) and use the template in Table 3 to identify ways in which they can be used to promote learning and the associated assessment considerations.

4.2 How can the effectiveness of RBL be evaluated?

The most direct way of assessing the effectiveness of RBL is to compare the extent to which changes have occurred in achieving learning outcomes before and after the introduction of RBL to a course. For example, the first year Child Psychology course at the University of Derby went over to an RBL format in 1995 using videotapes of the lectures given the previous year, study guides and surgeries. Statistical analysis of course work and exam performance showed no difference in results achieved by students using video compared to traditional delivery modes (McGhee, 1997). However, this is often not possible because changes may also have occurred in the learning outcomes, or in the methods of assessment, or both (Gibbs *et al.*, 1994a). Indirect ways of assessing effectiveness include:

- students' and tutors' subjective judgments of effectiveness;

- overall performance in the end of semester/year examinations;

- the percentage of students proceeding to the next stage;

- comparisons of the performance of individual students on RBL taught modules and conventionally taught modules;

- comparisons of the performance of students on those components of a module taught through RBL with their performance on other aspects of the module (Box 17).

Box 17:   The introduction of RBL can result in improved examination performance

Study packs have been introduced at Kingston University into several modules (Section 9.2). In the level I modules in Physical Geography and Environmental Science comparisons have been made of the performance of students in the end of module multiple choice examination on questions set on the material covered in the study packs with questions set on the material covered in the lectures. In both modules in 1995/96 the percentage of correct responses on the RBL questions was higher than for the lecture questions (Figure 4). The average number of correct responses on the RBL questions in Physical Geography was 57% compared with 43% for the lecture questions. The figures for Environmental Science were 60% and 44% respectively. This repeated a similar difference in performance the previous year.

Further information:
Susan Watts, School of Geography, Kingston University
email: S.Watts@kingston.ac.uk

Figure 4:   *The performance of students on RBL and lecture-based questions*

Environmental science I

Module GG1011A
Multiple choice questions (all compulsory)

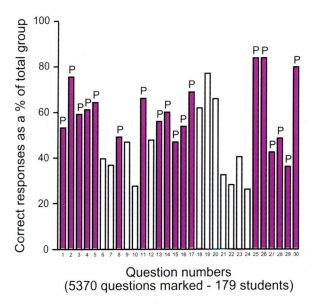

Question numbers
(5370 questions marked - 179 students)

Physical geography I

Module GG1001A
Multiple choice questions (all compulsory)

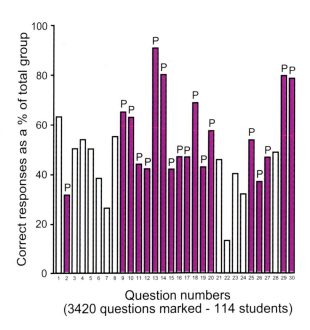

Question numbers
(3420 questions marked - 114 students)

P— Questions based on material covered in a resource pack

5 How can RBL be used to enhance teaching and learning?

Geography courses are adopting a widening range of teaching and learning methods, as is illustrated in the case studies in the other GDN Guides. A common feature of many of them is that they use predominantly paper-based materials to support independent learning or to replace or supplement teaching and the library. The use of other media, particularly electronic, is expanding, but apart from the use of some packages produced by consortia (for example, *GeographyCal*) or commercially (for example, some GIS packages), their widespread development in geography departments has up to now been constrained by the initial costs of software and hardware and the relative lack of expertise (see also Section 6.4). Although developments in the ease of authoring for the WWW is reducing these constraints. In this section I examine the need to pay more attention to the effective use of learning time and the potential for RBL to contribute to this both in-class and out-of-class. I go on to discuss the ways in which RBL may be used to substitute or supplement teacher-centred activities and other specific learning activities.

5.1 Why is there a need for an improved use of learning time?

Students place a high value on contact with staff, but the amount and form vary significantly between courses. In distance courses face-to-face contact is often minimal (Box 18 overleaf). Even in the more conventional campus-based courses most student learning already takes place outside the 'classroom' (including laboratory, studio and field-based classes) and the proportion of a student's time spent outside the 'classroom' is likely to increase in the future. It is therefore important that the limited time available for contact between teaching staff and students is treated as quality time and is designed to make the most of the interaction, while more support and structure is given to the non-contact time (Box 19). The design of modules and course units will increasingly be about planning what students do, rather than what teachers do, because each contact hour has to support an increase in out-of-class learning hours. Relatively more class contact will be used to set up, model and debrief out-of-class learning activities rather than present information (Box 20).

Box 18:   In distance courses face-to-face contact with tutors is often minimal

The Local Policy part-time course at Cheltenham and Gloucester College of Higher Education is largely delivered through the provision of interactive course books and project-based assignments. Students have two residential weekend schools each year at each level and most, though not all, Level 1 students have a monthly tutorial. Contact is maintained through telephone calls, monthly newsletters and written comment on draft and final assignments.

Further information:
  Elizabeth Skinner, School of the Environment, Cheltenham and Gloucester College of Higher Education; email: ESkinner@chelt.ac.uk

Reference:
  Skinner (1997)

Box 19:   Integrating a resource-based research project into a course can provide a structured in-depth learning experience for the whole class

The major assignment in the Graduate Seminar in Applied Human Geography at Southwest Texas State University in 1998 was a semester long project concerned with researching the information to enable an Ethnic Atlas of Texas to be compiled. The Graduate students workload singly, in pairs and small groups to gather data on one immigrant group important in the settlement of Texas. They used a wide variety of resources including census information, church records, city directories, archival newspapers, and tax records to produce an original report. The students also mapped census information for their ethnic group using 'Mapitude', which maps variables onto a base map of tract or block data for the entire USA. They also conducted interviews with people from their particular ethnic group to clarify their findings and add a humanistic touch to their work. Results from the research project are being compiled by the Department cartography staff to publish an ethnic atlas of Texas to celebrate the bicentennial of the University in 1999.

Further information:
  Susan Hardwick, Department of Geography and Planning, Southwest Texas State University; email: sh19@swt.edu

Box 20:   Integrating RBL into courses is about designing what students do in and out-of-class rather than what teachers do

The first part of the Environmental Philosophy third year module at Oxford Brookes University is designed around an exercise using a range of resources, including two handbooks, readings, and an extract from a classic film — *The Grapes of Wrath*. The exercise introduces the students to different ways in which geographers in the twentieth century have interpreted the relationships between society and environment. The two handbooks are used instead of lectures. The first outlines four environmental philosophies — environmental causation, cognitive behaviouralism, Marxism and Gaianism. This handbook

(19 pages) is given out at a preliminary meeting when the students are told they need to be familiar with all four philosophies by the day school the following week.

The module attracts approximately 70-80 students. At the beginning of the day school, run by four staff, the students are put into seminar groups of 16-20 students. The students then divide themselves into four groups and each group is randomly allocated one of the four environmental philosophies, along with some additional reading material. The groups are given 30 minutes to prepare a 5-6 minute presentation summarizing their specific philosophy and commenting on its strengths. A discussion follows in which other groups explore the weaknesses in what they have heard.

The seminar groups reconvene and the second handbook is distributed, which contains a selection of writings about the Dust Bowl years of the 1930s, some background notes on *The Grapes of Wrath*, and a brief content analysis of the seven scenes that comprise the 20 minute extract they will see. The groups are then given an hour to prepare a 7-10 minute presentation, in which it is suggested that they should reiterate their philosophical approach, give their interpretation of the Dust Bowl from that position, and pinpoint where, from that perspective, the 'blame' for the Dust Bowl disaster lies. The students have a second opportunity to see the film extract. They make their presentations following lunch. These are then assessed by the tutors and their peers against pre-circulated criteria, which are also used to structure a discussion, which seeks to pull the ideas together.

The experience of participation in the exercise acquaints students with the role that they play in later sessions of the module, namely, mastering a viewpoint by having to represent and explain its distinctive features to others. The exercise supports 18-24 hours of learning over a three to four week period. The first handbook took about 20 hours to produce, the second about 10 hours. The course has run for five years. In 1998 the exercise was modified for use in a new first year module.

Further information:
John Gold, George Revill and Martin Haigh, Geography Department, Oxford Brookes University; email: jrgold@brookes.ac.uk

Reference:
Gold *et al.* (1996)

RBL may be used to enhance the quality of both the contact time between staff and students and the non-contact time. However, the distinction is fairly arbitrary, because often a substantial part of an exercise started in class is designed for completion outside class time and many exercises are designed to substitute a form of independent learning for what was previously passive learning in class. Nevertheless, the distinction between the use in-class or out-of-class is a useful one to structure a discussion of the range of ways in which RBL may be used to enhance teaching and learning. A further distinction may be made between the use of RBL as a substitute or supplement for teacher-centred activities and as a substitute or supplement for specific learning activities.

Activity 12

- Take a module/unit that you teach and calculate how many hours of contact your students have with you and how many out-of-class hours of learning these are meant to support (a quick way to estimate the total number of learning hours per module/unit is to divide the number of modules/units a full-time student takes in the semester/year into 40 hours, an assumed full-time working week, and multiply by the number of weeks in the semester/year. By subtracting the contact hours from this total you have the number of out-of-class hours).

- Do you make the expectation explicit to your students that for every one hour they spend in your class they should be spending x hours undertaking assignments, doing the essential reading, reviewing and revising the course material and so on?

- Do you allow sufficient time for students to undertake these out-of-class activities?

5.2   How can resources be used in class?

Formal contact between geography staff and students takes many forms including lectures, seminars, tutorials, practicals, laboratories, fieldwork and dissertation supervision. There are many examples of the use of RBL by tutors to enhance the contact time in these circumstances in the other relevant Guides in the GDN series (Agnew & Elton, 1998; Birnie & Mason O'Connor, 1998; Clark & Wareham, 1998; Livingstone *et al.*, 1998). RBL has been used extensively for decades in practical classes, particularly in techniques classes, but is increasingly being incorporated into other classes. This may involve an exercise integrated into, say, a lecture or tutorial, where students, for example, analyze and discuss some data, a case study, a video, or an audio-tape (Boxes 12, 21 and 22). In other cases the whole session may be based around an exercise or series of exercises using a variety of information resources.

Box 21:   Careful planning is needed when using videos to ensure that they are an effective learning resource rather than just an entertainment medium

In the Level 2 Urban Hydrology module at the University of Leeds a video on *Managing the Humber Estuary: the Environment Agency's Approach* (Environment Agency, 1997) is shown. It has the merit of being short (17 minutes) and the Catchment Management Plan for the Humber Estuary is available in the Library. The session fits within a pair of lectures that explicitly look at case examples of issues in the Aire, Calder, Don and Rother catchments. Amongst the issues addressed are inter-catchment water transfers and the disposal of Birmingham's treated water that is sourced in the Welsh Severn catchments, but is treated and returned to the River Tame and so reaches the sea via the Humber.

Before starting the video the students are asked to make notes under three headings:

- The role and responsibilities of the Environment Agency.

- The factors that impact on water quality in the Humber.

- Implications for wildlife around the estuary.

After the video they are given 5 minutes to collaborate with their neighbour to tidy up their notes and make sure they have covered all the points. A brainstorm then takes place using the above three headings. Most of the class have never visited the Humber area, and seem quite amazed by it. The extent of the Hull, Grimsby and Scunthorpe dock and industrial areas is greater than they expect. The scale of the issues, volumes of water and quality to be treated, become slightly more real than in the tables in the Catchment Management Plan.

The video replaces the section of the lecture on the role of the Environment Agency and its links to industry, local communities and environmental groups. The video is useful because it makes many points in a short time frame. Seeing the landscape reinforces awareness of the flooding vulnerability of the Humberside estuary and the multiple impacts the estuary has to carry. The video shows the Mablethorpe beach and holiday trade alongside wetland nesting sites, fisheries, industry and port works and makes the point that the Humber drains about 20% of the UK, a land area which is home to a very large population.

Further information:
   Pauline Kneale, School of Geography, University of Leeds
   email: P.Kneale@geog.leeds.ac.uk

Box 22:   RBL exercises may be integrated into lectures and tutorials

Active learning exercises can help students understand the content and concepts of a course

In her 201 Human Geography course Sarah Bednarz has devised nine activities which relate to the ten topics covered in the course. The exercises range from analyzing the diffusion of the granting of presidential voting rights to women in the States; to researching the predominant ethnic groups, languages, and religious groups in 25 different nations; to interviewing a student who is a native of a foreign country. The students are encouraged to form study groups with three or four other students, but have to write up assessed work individually. The exercises include practicing various skills, including drawing maps and graphs, interpreting and analyzing data, searching for information, interviewing and group work. Each of the activities is assessed and together they count for 34 per cent of the student's grade.

Further information:
   Sarah Bednarz, Department of Geography, Texas A&M University
   email: s-bednarz@tamu.edu

Graphs and remote sensing images may be used to encourage tutorial discussion

In a tutorial class at King's College London, groups of first year students are presented with either a series of graphs with their axis labels erased, or a series of remote sensing images. They are then asked to devise labels for the

(cont.)

---

graphs or develop a discussion of the images. Each statement they make has to be justified with evidence from the resource examined. The students have 10-15 minutes to complete the exercise, usually without the tutor present. After which they have to present their conclusions with respect to each graph/ image. The exercise encourages some intensive discussion and interaction between the students.

Further information:
   Mark Mulligan, Department of Geography, King's College London
   email: mark.mulligan@kcl.ac.uk

Reference:
   http://www.chelt.ac.uk/gdn/abstracts/a4.htm

Activity 13

   • List the variety of ways that you use *resources* currently in your lectures and tutorials.

   • Could you make more effective use of resources to encourage active learning?

5.3 How can RBL be used to structure and support out-of-class activities?

The problem with reducing student contact time with teaching staff is that it can simply result in reduced student effort. To avoid this effect there is a need to provide greater structuring of the non-contact time around meeting learning outcomes by, for example:

   • providing clear specification of independent learning tasks, such as structured projects and directed reading (Box 23);

   • giving good access to, or provision of, learning resource materials, for instance through readers and resource packs, and resource centres (Boxes 14, 49, 50 and 53);

   • increasing student responsibility for learning in forms such as self and peer assessment (especially for feedback), peer tutoring, and tutorless seminars (Box 47).

RBL may be involved in all of these cases. Learning guides may be used in the first and last examples to give students advice on how to undertake the activity, while the substantive content of the course may be conveyed through RBL in the second example. A field study guide could be designed to meet the first two objectives, while a dissertation guide is likely to focus on the processes of undertaking the dissertation and various administrative and presentation details (Livingstone *et al.*, 1998). Where developments such as these take place alongside lectures, seminars and laboratories, they are often justified on the basis of maintenance or enhancement of quality.

Box 23:   A clearly written guide on the task to be undertaken and advice on how it may be completed is an important pre-requisite for successful project work

The main piece of coursework in the second year module 'The Geography of Global Issues: Environmental and Demographic Aspects' at St. Mark and St. John University College consists of a group project examining 'The Challenge of World Health'. It counts for 50% of the module assessment. The precise topic of the project changes each year to avoid work re-circulating. In 1997/8 the project involved groups of three or four students taking the role of a commercial research agency producing a report on HIV/AIDS for the Kenyan government. In previous years the project has focused on child mortality and fertility trends. Ninety-five students took the module in 1997/8 generating 25 group projects. Each project took about 40 minutes to mark giving a significant saving in assessment time compared with marking individual project work.

The module tutor produced a short booklet (24 pages), which introduces the project, lists key references and gives the assessment criteria. About half the booklet is concerned with working effectively in a group, and giving proformas to record meetings, self-assessment of the contribution to the project and a group-proposed allocation of marks. Two short articles are included in the booklet; one deals with 'The Geography of HIV/Aids' (Digby, 1997) and the other with 'Why work in groups?' (Vujakovic *et al.,* 1994). Links to relevant WWW sites are provided on the College Web site. Emphasis is put on the students analyzing and presenting data using EXCEL and MAP91.

Half of the 24 hours class contact time is allocated to the project. This consists of two one-hour lectures which introduce concepts, and five two-hour seminars in which the tutor and/or a technician is available to support individual groups. The time spent with each group varies according to demand, but it is focused teaching aimed at encouraging deep learning. It is estimated that the students spend at least another 30 hours outside the classroom working on the project.

Further information:

Sue Burkill, Department of Geography, St. Mark and St. John University College; email: stasmb@lib.marjon.ac.uk

Activity 14

- What do you do currently to help your students *learn effectively out-of-class*?

- How can you improve on the common practice of giving students a reading list and telling them you will expect them to show knowledge and understanding of the material in the end of semester/year examination?

- What else could you do to support, structure and encourage student learning?

- Write down your suggestions.

5.4   How can RBL be used as a substitute or supplement for teacher-centred activities?

Whether used in-class or out-of-class, a common application of RBL is as a supplement to or replacement of teacher-centred activities. By teacher-centred activities I mean occasions when tutors are presenting the content of courses or giving guidance and advice on undertaking assignments in a face-to-face situation, such as in a lecture, tutorial or practical class.

Content and advice may be substituted for:

- directly, by providing, for example, a set of word processed lecture notes; an assignment guide; or an Open University-standard illustrated package;

- indirectly, by providing, for example, an independent learning assignment; a guided reading exercise; or a reader covering the same topic.

Hybrid approaches are often the most effective, such as a tutorial-in-print, which provides content along with in-text questions, activities and commentary; or a course study guide, which, as well as introducing the course, provides hints and ideas on how to obtain the most out of the course and helps students to organize and pace themselves (Box 24).

Box 24:   The greater the emphasis on independent learning the more important is a clear course study guide

The Open University second-level geography course, *The Shape of the World: Explorations in Human Geography* (D215), is based around a set of 5 text books, which include chapters written by the course team, short readings and activities, 8 audio programmes and 12 television programmes. There is a separate study guide for each text book and associated audio and TV programmes and tutor marked assignments. The Course Guide (15 pages) covers the following:

1	What's the course about?	2
2	Why a geography course?	4
3	How is the course organized?	5
	3.1 The five volumes	5
	3.2 The significance of "Chapter 6"	7
	3.3 Integrated Readings	10
	3.4 Study Guides	10
	3.5 The role of television and audiocassettes	10
4	What assessment is used?	11
5	Pacing your studies and the role of "review weeks"	11
6	"Using" course materials	11
7	Tutorial strategy and local experience	13
8	The place of D215 in your OU studies	14
	Acknowledgements	15

Probably the most useful item in the Course Guide is the Course Plan which gives a weekly timetable of activities to help students pace themselves (Figure 5). Monthly tutorials arranged by the Open University Regional Offices are available to assist the students although the emphasis is on distance learning.

Reference:
    Brook (1995); see also Box 45

Figure 5: *D215 Course Plan*

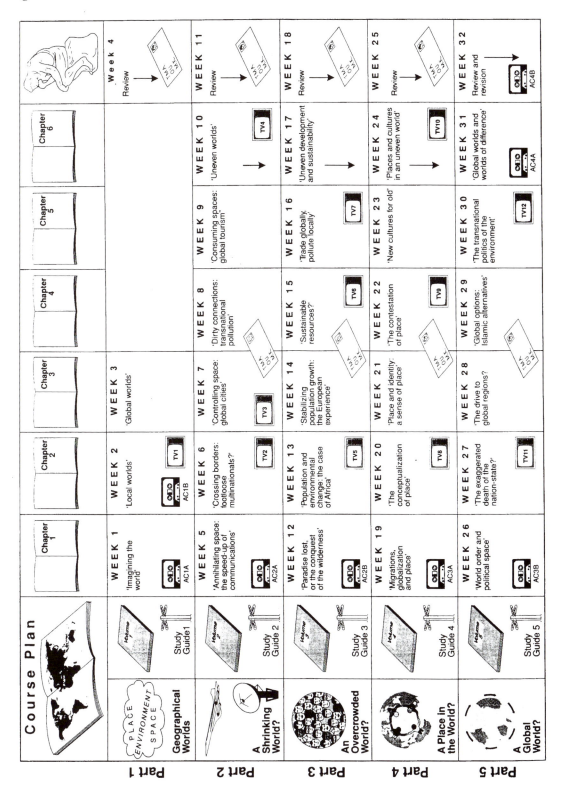

The aims of developing and using substitute materials for teacher-centred activities may include:

- reducing teaching costs, especially where sessions are repeated;

- overcoming room shortages;

- providing specialist expertise which would not otherwise be available;

- ensuring that students have good notes to support out-of-class learning activities;

- avoiding student passivity and encouraging independent learning.

(FDTL National Co-ordination Team, 1998a).

Complete substitution of face-to-face contact by RBL is rare and is usually inappropriate (see Section 8). In many cases materials are used as supplements rather than replacements. Thus lectures may still be given, but students are encouraged to listen and take part in activities rather than take notes (Box 25).

Box 25:   Making lecture notes available to students enables them to concentrate more on the lectures and/or enables some or all of the time spent in lectures to be used for more active learning

Trials have been taking place at Exeter University with the use of *paperless lectures*. Electronic presentations are used in lectures in various physical geography and statistics courses. Full lecture notes are prepared using the Microsoft PowerPoint software and are presented from a PC or laptop computer via an overhead projector and tablet, or a data-projector, to classes ranging in size from less than 20 to more than 100. The electronic presentation for each lecture is made available for student revisiting by placing the files on a server to which there is access from the teaching clusters of PCs. Students are instructed in the use of the PowerPoint viewer software and can review or print off the lecture notes at any stage after the material has been presented in class.

The approach has the following advantages:

- It is relatively easy for staff to produce a set of lecture notes which are professional in appearance and attractive to read.

- Presentations can be updated with new material easily and with little lead time.

- Students can focus on what is being said in lectures rather than on the process of note-taking.

- There is a learning resource produced which provides backup for the student body.

- Toggling between a PowerPoint presentation and statistics software (such as MINITAB), during classes or when students are revisiting, facilitates practical instruction of statistical techniques.

- Costs in producing lecture notes are significantly reduced.

The reaction of students to this approach has been favourable, and the range of courses presented in this manner is being extended.

Further information:
   Bruce Webb, Department of Geography, University of Exeter
   email: B.W.Webb@exeter.ac.uk

Reference:
   http://www.chelt.ac.uk/gdn/abstracts/a70.htm

If the content or advice is available in a written (Box 26), audio (Box 15), or video (Box 27) form then the amount of face-to-face contact may be reduced and/or a different, more effective form of learning may be substituted. The role of the lecturer then shifts away from being the provider of content to guiding the students through learning activities and supplying feedback (see also Agnew & Elton, 1998; Healey, 1991; and Section 9.4).

Box 26:   Where lectures have been replaced by RBL packs some of the time saved can be used to provide small group teaching

The year one geography modules at Oxford Brookes are taught through the use of written course units which the students have to study themselves. The lecture time saved is then invested in providing weekly seminar groups, in which progress is reviewed and specific exercises in support of the written materials are carried through. The course units are designed as workbooks which specify the tasks the students need to undertake. Many of them synthesize published interpretations of the issue under study and/or provide a selection of articles or extracts from books. A programme of videos are an integral part of the work of particular units.

Further information:
   Simon Carr, Geography Department, Oxford Brookes University
   email: sjcarr@brookes.ac.uk

Box 27:   Lecturers who have had hands-on experience of making a video of their own lectures have generally enthused about the potential

Videos of lectures gives students access to lectures outside the tyranny of the lecture slot, which can increase access for part-time students and students with timetable clashes. For staff the main advantage is that they can avoid repeating lectures. However, until recently producing a video has been relatively expensive and is often seen as an imposition by the lecturers. The strength of these objections has been substantially reduced at the University of Derbyshire through using VESOL, Video autoEditing System for Open Learning invented by Chris O'Hagan, which gives the lecturer control over the process.

(cont.)

This simple system consists of a set of four or five fixed cameras giving say a wide angle view of the front of the lecture room, a close up view of the lecturer standing at the lectern, a view of the OHP screen, and a view of the whiteboard. The lecturer selects which camera view is appropriate from a control panel at the lectern. There is also a small monitor at the lectern and another wall mounted one so that the lecturer can always see what is being recorded. The system costs c£10,000 and lecturers are competent to use it following an hour's training. A cheaper system has also been developed to go in a smaller room which allows the videoing of presentations of varying lengths without an audience being present.

The videos can be viewed straightaway without any editing, as a recording of the lecture or extracts may be played perhaps to promote discussion in a tutorial. With editing a series of mini lectures can be created or several views from different experts can be included on a single tape. By making the videos available in a resource centre students from other disciplines can also view them and explore the interfaces between subject areas.

Further information:
    Chris O'Hagan, Educational Methods and Media, University of Derbyshire
    email: c.m.ohagan@derby.ac.uk

Reference:
    O'Hagan (1995)

Activity 15

"A lecture is the process by means of which information is transferred from the note-book of the lecturer to the note-book of the student without passing through the minds of either"

(Thomas Huxley).

Imagine that your students will have to learn about the topic of your next lecture without you lecturing in person.

- What alternative ways could this be accomplished?

- Which of these involve some form of RBL?

- How might you provide support, structure and encouragement to help your students achieve the objectives of the session?

- How might the time your students spend with you in the lecture theatre be used most effectively?

---

* *Lectures do, nevertheless, provide an important role in learning in HE. They are a reasonably efficient way of transmitting information. However, students listening or taking notes while a lecturer speaks are unlikely to achieve higher level educational objectives, such as critical thinking and changes in attitude. For these objectives to be attained in the lecture theatre requires a more reflective style of lecturing, with the lecturer acting as a guide and facilitator and providing active learning opportunities integrated into the lecture (Agnew & Elton, 1998).*

5.5 How can RBL be used as a substitute or supplement for specific learning activities?

As well as substituting for some teacher-centred activities, RBL may be used as an alternative to some student-centred activities which are difficult to provide, staff or resource. Usually the learning resource involves detailed instructions for activities as well as reading material, and replaces or is an alternative to an experiment, a fieldtrip, or other first hand experience (Box 28). The learning activity is usually only one part of a course. Students with perceptual or motor restrictions may also gain access to learning activities and experiences that would otherwise be unavailable to them (FDTL National Co-ordination Team, 1998a).

Box 28:   Field trails can provide field experience to students without a member of staff being present

Three field trails were produced by the Department of Geography, Lancaster University to enable first year tutorial groups to gain field experience on their own and replace the previous staff led coach trips. Each trail is designed to give a days work to a group of six to seven students. Two of the trails give students six or seven alternative sites to examine to avoid over-saturating particular places. The trails have been published as A5 booklets (each c52-56 pages).

They are structured around a set of common headings:

- Notes on completing the trail (aims, preparation, field safety, how to follow, post-trail).

- Background information.

- Trail.

- Answers section (questions and room to write answers).

- Selected references.

Further information:
Gordon Clark, Department of Geography, University of Lancaster
email: g.clark@lancaster.ac.uk

Reference:
Higgitt & Higgitt (1993a, b and c)

Audio-visual resource are beginning to be used to supplement or replace laboratory experiments and fieldcourses (Box 29). McKendrick & Bowden (1997) found that in a survey of 65 geography departments in the UK, 24 per cent had videos offering instruction in fieldwork techniques and 15 per cent had them for instruction in laboratory research techniques. A higher proportion (38 per cent) used videos to introduce students to fieldwork locations.

Box 29:   Videos can increase the opportunities for students to experience laboratory techniques and save staff time

Videos are used at Cheltenham and Gloucester College of Higher Education to demonstrate laboratory techniques. They are used for a variety of purposes to:

- demonstrate frequently used techniques;

- illustrate the design and outcome of experiments which use equipment the department does not possess or is not available for undergraduates to use;

- show experiments which take a long time to prepare or undertake;

- illustrate experiments which use expensive materials or are potentially hazardous;

- introduce students without a scientific background to the laboratory;

- give access to laboratory-based learning experiences for students with motor restrictions;

- enable students to revise techniques, particularly for use in their dissertations.

Further information:
Jacky Birnie and Carolyn Roberts, School of the Environment, Cheltenham and Gloucester College of Higher Education; email: jbirnie@chelt.ac.uk; crroberts@chelt.ac.uk

Computer-based resources are also beginning to be developed to supplement fieldtrips and to provide access to a range of places which students cannot afford to visit. Several virtual fieldtrips have been developed over the last few years. Many are little more than guided tours illustrated with slides. However, they can provide a valuable resource, where they are integrated into courses with clear objectives and exercises which make use of the educational potential of these tours. More elaborate virtual fieldcourses are making use of the potential of computers to hold large databases which can be analyzed using GIS techniques (Box 30). Further details on computer assisted learning packages, which deal with fieldwork techniques, and virtual fieldwork WWW sites are given in the GDN guides dealing with fieldwork and ICT (Livingstone *et al.*, 1998; Shepherd, 1998).

Box 30:   The greatest potential of virtual fieldwork is to enhance work in the field rather than replace it

The Virtual Field Course (VFC) project based at Birkbeck College, the University of Leicester and Oxford Brookes University aims to recreate the teaching traditionally executed in the field in a virtual computer world. The team's aim is to develop a tool kit which others can use to visualize and analyse their own data. By superimposing different maps in a geographical information system students can, for example, analyse the associations between geology, landforms, soil and vegetation or they can compare maps showing the same area at different dates. They may also compare data they have collected in the field with secondary data, such as census and historical maps, held in the VFC software. Exemplar VFCs are being developed for popular fieldcourse locations in the UK, including Dartmoor and the North Norfolk coast and for remote locations in Bolivia and Namibia.

Further information:
  David Unwin, Department of Geography, Birkbeck College University of London; email: D.UNWIN@geog.bbk.ac.uk
  http://www.geog.le.ac.uk/vfc

References:
  Moore (1998); Unwin (1998)

Activity 16

- What learning experiences are difficult to provide, staff or resource in your department?

- Could RBL be an alternative for any of these?

- Think of one way in which, say, designing a learning exercise, or making a video, or using a computer assisted learning package, could provide your students with a learning experience which they would otherwise miss out on.

6 What are the strategies for embarking on RBL?

An increasing number of institutions are adopting a mixed-mode approach to the use of resource-based learning materials, in which independent learning packages are being used by on-campus students alongside conventional teaching methods (Table 4). These materials were sometimes originally developed for distance learners; in other cases they have been developed to give greater flexibility for campus-based students. In this section I begin by examining some of the strategies which individuals and departments may adopt to introduce RBL into their courses. I go on to comment briefly on integrating RBL with more conventional teaching methods. The remaining parts of the section are concerned with a discussion of the cost effectiveness of RBL, how the introduction of RBL packages can be costed, and what strategies may be followed by individuals and departments to make RBL cost effective.

Table 4: *Institutional strategies towards open learning*

Approach	Features
Conventional	Campus focused; based on face-to-face activities; students attend at predetermined locations and times; considerable infrastructure requirements; little or no use of open learning materials
Distance Learning	Teachers and students separated by distance and time (open learning materials used over several years); courses primarily or exclusively based on open-learning packages
Dual-mode	Separate programmes using conventional approaches and distance methodologies with little synergy between the two
Mixed-mode	Open learning materials integrated into campus based provision

Source: Based on information in Fellows (1997, pp.147-148)

6.1 How can I and my department introduce and develop the use of RBL?

Although course guides and class handouts are almost universally used by lecturing staff in geography departments, other forms of RBL, such as in-lecture learning exercises, independent learning packages, readers, audio-video learning resources, and CAL packages, are more selectively used. A sensible strategy to follow, if these latter forms of RBL are new to you or to your department, is to experiment and evaluate the experience on the 'think big, start small' principle (Box 31). For an individual this might involve replacing a lecture or a block of lectures with an open learning package, or introducing a RBL exercise into a lecture, or making a video to demonstrate a piece of equipment. For a department, several staff could be encouraged to experiment in these ways with different classes so that a wide range of experiences can be evaluated quickly, or a new module might be designed from the start to incorporate a significant amount of RBL. By starting small, RBL may be introduced gradually in a series of stages each building on the previous one and incorporating additional features as experience is gained (Gibbs *et al.*, 1994a).

Box 31:   Phasing-in the development and use of RBL packages can be helpful to both staff and students

Problem

Making the initial transition into comprehensive resource-based learning can present difficulties for course co-ordinators and students alike. It takes time and thought to develop the sort of resources which are really robust enough to stand alone.

A possible solution

One way of coping with the transition is to provide resource packages alongside the more traditional lecture-based course, so that students who do not absorb material well in lectures have a chance to review the material on their own. The resource packs do not have to be complex. The course guide in the Stage 1 geography papers at the University of Auckland includes a weekly study guide to the resources available. It briefly outlines the theme for the week, provides a list of readings, and, most importantly, gives a series of keywords which relate to the theme. The definition of the terms may be found in the prescribed texts, or in the Dictionaries of Human/ Physical Geography. Elsewhere in the guide there is an outline of the laboratory for the week, with any background material which would be useful to have to hand before the lab. All of the readings are in the reserve section of the library, so all students have reasonable access.

When students fail to keep up it is often because they miss a section through illness or other reasons, or when they fail to understand a part of the material and do not move on beyond that. Key words have proved to be a powerful way of getting around this problem and the students are encourage not only to learn the definitions but to look at how, when, where, what and why that term may apply or be applied. Once they have worked with the vocabulary their understanding of the texts and lectures improves immensely, and less students give up. It also has the advantage that students who don't have good study skills when they start are provided with a fairly mechanical way of initially getting into the material, and developing good habits.

Resource implications

The guides require quite a lot of work before the course starts, both to get them written and printed, and to make sure that the resources are available in library collections. It means that lectures have to be written, or at least planned in some detail before the course starts, rather than as required, but they fortunate

(cont.)

---

in having a Senior Tutor attached to the first year programme who co-ordinates all the academic tasks associated with the course. The development of the course material is part of her role.

This system has put the responsibility for understanding the material into the student's hands and greatly reduced the number of students seeking assistance with the obvious reduction in the time spent by academic staff with individual students.

Further information:
    Margaret Goldstone, Geography Department, University of Auckland
    email: m.goldstone@auckland.ac.nz

Reference:
    GDN WWW Resource Database, includes an example page from the
    guide, http://www.chelt.ac.uk/gdn/abstracts/a85_app.htm

Activity 17

- Write down two ways in which you could *integrate an aspect of RBL* which you haven't tried before into one of your modules or units.

- Select forms of RBL which would enhance the quality of your students' learning, and which are under your control to introduce.

- Plan what you will need to do to introduce at least one of these ideas the next time you run this module or unit.

Departments wishing to develop a strategy for the development of RBL need to address a range of issues. Many of these are discussed elsewhere in this Guide. They include:

- What is the strategy trying to achieve? (See e.g. Case Studies 9.2 and 9.5)

- How will RBL be used to enhance learning? (See Section 5)

- How can the introduction of RBL be costed and be made cost effective? (See Sections 6.4 and 6.5)

- Which courses in which years are going to be targeted? How will they be chosen?

- How will the materials be designed to promote learning? (Section 6)

- Who will prepare and coordinate the materials? How will this process be facilitated? Will specialist help be bought in (e.g. to help make the text interactive)?

- What support will the students using the materials need? (See Section 8.1)

- What support do staff need in developing and using RBL? (See Section 8.2)

- How will learning support staff be involved? (See Section 8.2.1)

6.2   How can RBL be integrated effectively with other learning methods?

In a conventional university few modules are taught entirely through independent learning packages. Even in the Open University there are face-to-face or telephone tutorials arranged, though not all students avail themselves of the opportunities (see Box 24). RBL needs to be seen by both staff and students as just another learning method to be used alongside other methods of learning and teaching as and when it is appropriate (Box 32).

Box 32:   RBL is usually best integrated with other learning methods

As part of the Hertfordshire Integrated Learning Project (HILP) the course team have developed a problem-based case study focused on the Norfolk Broadlands. This combines classroom study, through lectures, group workshops and skills workshops, with self-directed study, using paper-based and computer-based learning resources (Figure 6).

HILP is an institution-wide project supported by the FDTL. Staff from eleven disciplines, including environmental sciences, geography and geology, are currently involved in the project. HILP has as its focus the integration of skills development into the academic curriculum with the preferred vehicle for this integration being problem-based learning. As part of this work, the core team are developing a set of transdisciplinary problem-based case studies to be used by the disciplines involved with the project. The first of these case studies is focused on Broadlands, East Anglia and has been developed in collaboration with officers from the Broads Authority. It is designed so that different disciplines can focus on particular aspects of Broadland. For example, the environmental chemists concentrate on questions of water quality, whilst the lawyers concern themselves with navigation rights.

As part of a second-year module (1997/98) on 'Global Change', the geographers and environmental scientists undertook a seven week case study to investigate and make proposals for sustainably developing the Upper Waveney Valley for recreation/tourism and environmental enhancement. Sixty-three students, working in groups of three or four, were allocated a particular site which they investigated on a field visit. Their task was to produce a strategic plan for their site and present it in the form of a poster at the end of the case study. A programme of lectures and skill workshops provided the case study with a framework. The following *resources* were made available:

- A student briefing pack: a document giving the aims, structure and timetable of the case study, guidelines for the poster presentation, assessment criteria, and the method used for distributing the group mark between individuals.

- A consultancy brief from the Broads Authority: a two page document giving the background to the proposal for a Tourism and Environmental Enhancement Plan; the aims of the plan; a checklist of points for the students to consider when making their site- specific proposals; and a list of key valley features giving an overview of the valley together with site specific details.

- A video of a lecture given by the Broads Authority officials who collaborated on the project.

(cont.)

- A Broadland Reader: a collection of extracts from various books and reports covering landscape, environmental and socioeconomic issues in the River Waveney valley and surrounding area.

- A Sustainable Development Reader: a collection of articles giving general information on sustainable development together with a section on sustainable tourism.

Even with a shelf life of three years, preparation of resources for this case study was costly in terms of staff time. However, because of the multidisciplinary nature of HILP, the utilization of the material by more than one discipline (e.g. 300 students in 1997/98 from Environmental Sciences, Law, Music and Business Studies) makes the exercise more cost efficient.

Further information:
   Jennifer Blumhof, Marianne Hall and Andrew Honeybone, Division of Environmental Sciences, University of Hertfordshire
   email: J.R.Blumhof@herts.ac.uk;

Reference:
   http://www.cs.herts.ac.uk/hilp/

6.3   How cost effective is RBL?

The cost effectiveness of different modes of learning and teaching is central to any learning and teaching strategy. Cost effectiveness requires both costs and benefits to be considered together (Hunt & Clarke, 1997). However, most attempts to quantify cost effectiveness has focused on the cost side. Generalizing about the cost effectiveness of RBL is difficult. As we have seen, the forms which RBL take, and the ways in which it is used, vary widely (Box 1 and Figures 1 and 2). Moreover, the media used for RBL have a marked effect on the production costs. For example, the Open Learning Foundation estimated in the early 1980s that the cost in time for the development of learning packages was 50 hours for one hour of tape presentation, 200 hours for one hour of CAL, and 500 hours for one hour of video presentation. These figures ignore the cost of the capital equipment and the training time involved to use the technology. They compare with a figure of 10 hours which Baum *et al.* (1985) give for the preparation of a one hour lecture. Although developments in user-friendly technology and authoring programmes may have reduced the absolute times, the relative differences probably remain.*

As part of the Dearing Committee Report on HE, a comparison was made of the cost structures of different modes of teaching (NCIHE, 1997, Appendix 2). Although less generous in the estimates of the amount of staff time needed to support one hour of student learning, the relative differences are still large. The authors make an important distinction between RBL which is developed externally to the institution, such as textbooks, CDs and software, and RBL which is developed in-house, such as printed lecture notes, courseware programmes and customized spreadsheets (Table 5).

---

* The Open University has gained considerable expertise over more than 25 years of developing various forms of independent learning material. They allow two to three weeks for making a half-hour TV programme (video) (including reconnaissance, liaison with BBC, post-production) and 10 hours of academic time for one hour of student study time (including two to three drafts, fully discussed with colleagues) (Dee Edwards, 9 March 1998, personal communication).

Figure 6: Possible components of the Broadland case study

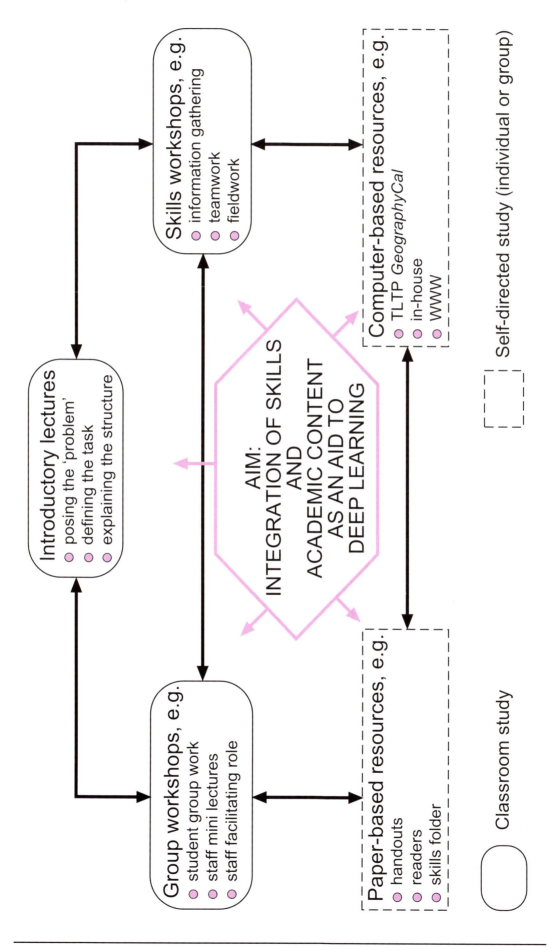

Table 5:   *Staff time to prepare material for one hour of student contact using different methods of teaching*

Lectures (up to 100 students)	3 hours preparation + 1 hour presentation = 4 hours
Small groups (up to 10 students)	1/2 hours preparation + 1 hour presentation = 1.5 hours
RBL (external)	2 hours preparation for use of existing resources
RBL (in-house)	20 hours for developing in-house.

Source: NCIHE (1997, Appendix 2)

The report recognizes that different forms of RBL vary widely in the time taken to develop them. Thus it suggests that two hours preparation time would be appropriate for selecting materials for students to read and work through, with a written study guide; twenty hours would be appropriate for developing materials with low production costs; and many IT-based developments would take 100-200 hours per student study hour. RBL preparation and development requires the same amount of time regardless of the number of students, whereas the costs of lectures and small groups rise in steps.

Using these assumptions the authors model the relative costs of different combinations of teaching methods under three different scenarios (Figure 7). In the 'Traditional' model small group teaching and lectures dominate. In the 'Current' model lectures and in-house RBL have expanded at the expense of small group teaching in response to the growth in student numbers. In both these scenarios a small amount of external RBL occurs, largely based on use of the library. A possible 'Future' model is also shown in which the amount of small group work is restored to a more acceptable level, and sees a major increase in the proportion of external-RBL, largely at the expense of lectures. The 'Current' model is successful in keeping costs within bounds as student numbers increase, but the reduced amount of group work diminishes the student experience. The advantage of the 'Future' model is that not only is a significant amount of small group work retained, but it also achieves a flatter and lower overall cost curve (i.e. the total cost for 100 students on the 'Future' course is similar to the total cost for 50 students on the 'Traditional' course).

The main problem with the 'Future' scenario is that, at present, the materials do not exist in anything like sufficient amounts for institutions to buy-in to support the envisaged amount of learning time. As the authors recognize, for RBL to cover the majority of student learning time the materials will need to be very well designed (whether print, audio visual, or software) and capable of supporting students working independently. For this to happen several preconditions are necessary:

- an acceptance by staff that they are prepared to use RBL materials which they have not designed themselves;

- an acceptance by the majority of students that they are prepared to double the proportion of their time working independently and that this time is filled using packages not developed in the institution in which they are studying;

Figure 7: *Alternative combinations of teaching methods (Source: based on data in NCIHE, 1997, Appendix 2)*

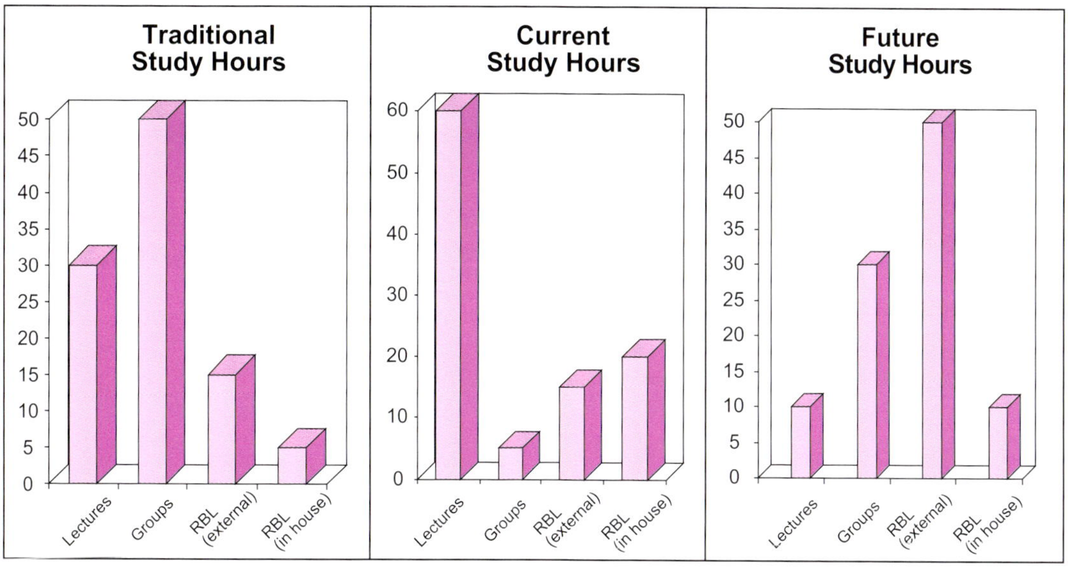

- the existence of a market needs to exist which is large enough to attract individuals and institutions (whether on their own or in consortia), publishers or other companies to produce the materials.

As yet, the market for RBL is immature, particularly on the computer-based front (Atherton, 1998). The most attractive topics are those which are taken by large numbers of students and are not quickly dated. These are limited in geography in comparison to many other subjects, such as economics or chemistry, because of the lack of agreement on what constitutes the core topics. In a post-modern world variety and difference between syllabi in different institutions, particularly in human geography, are often encouraged. The best potential seems to lie in introductory material, especially in physical geography, techniques, and inter-disciplinary topics which are relevant to several different disciplines.

Although I have doubts about whether the 'Future' model will work out in geography quite as envisaged in the Dearing Committee report, it is clear that the pressures facing HE are moving us in the direction of increasing the proportion of student learning hours covered by RBL. Continued development of in-house learning packages seems inevitable, particularly while few externally produced packages are available. Given the much smaller size of the 'market' within an institution, the development of in-house RBL packages therefore need to be made as cost effective as possible (see Section 6.5).

6.4 How can the introduction of RBL packages be costed?

Much RBL is introduced on the basis that it will lead to savings in cost or time. Whether this is true depends on the design of the RBL component or course and the design of the component or course that it is replacing. The figures that come out of a costing exercise are also very sensitive to the assumptions about the time taken to complete various tasks and the rates charged for teaching staff, accommodation, reprographics and so on. RBL is not automatically cheaper. Where the chosen form of the RBL costs the same or is only slightly more than a conventionally taught course, careful consideration needs to be given as to whether any gains in flexibility and improved educational effectiveness warrant its introduction. Undertaking a costing exercise is a prudent step to avoid making a commitment to an unsustainable course design decision (Box 33 and Activity 18).

Box 33:   Before embarking on large-scale investment in RBL it is essential to examine the costs

The example below raises a variety of issues about whether is would be sensible to adopt RBL. In this example, costs in year one are substanitally increased owing to the preparation and printing of materials, while in subsequent years the RBL course produces 14% savings. Over five years the RBL alternative costs almost exactly the same as the conventional course.

Costing RBL

This example costs a course for 120 students over 12 weeks. In the conventional course twice-weekly lectures were given to all 120 students, weekly seminars were given to groups of 12 students, and support was provided by infrequent tutorials. The RBL alternative uses learning packages to deliver the course in six two-week units; there are weekly workshops for 24 students, supported by regular surgery hours.

Conventional course	RBL course

Teaching

Conventional course	RBL course
2 hours lectures per week:	1 hour workshop per week:
24 hours @ £48 = £1152	60 hours @ £48 = £2880
1 hour seminar a week:	6 hours surgery time available per week:
120 hours @ £24 = £2880	72 hours @ £24 = £1728
48 hours tutorial time @ £48 = £2304	
Total teaching cost = £6336	Total teaching cost = £4608

Assessment

Two assignments and an exam: no change from current system

Accommodation

Conventional course	RBL course
24 hours lecture rooms @ £60 = £1440	60 hours large seminar rooms @ £20 = £1200
120 hours small seminar rooms @ £10 = 1200	72 hours small meeting rooms @ £10 = £720
48 hours offices @ £10 = £480	
Total accommodation cost = £3120	Total accommodation cost = £1920

Library use

Maintained at current levels by spreading range of reading

Reprographics

Conventional course	RBL course
About 50 handouts @ 2p:	6 40-page units @ £2.00 = £1440
£1 x 120 = £120	24 pages copyright material @ 9.6p = £276
Total reprographic cost = £120	Total reprographic cost = £1716

Preparation time

None over and above normal	240 hours writing (40 hours x 6 units)
expectations for lecturers to	@ £24 = £5760
update their courses, prepare	60 hours dtp time (10 hours x 6 units)
handouts, etc.	@ £10 = £600
Total preparation cost = £0	Total preparation cost = £6360

Costs in year 1

£9576	£14604

Costs in subsequent years

£9576	£8244

Total costs over five years

£47880	£47580

Notes:

1. It may be difficult to put a price on teaching hours or accommodation usage and so it may only be possible to count and compare hours rather than costs. There may be no saving in overall costs as a result of reducing lecturers' contact hours, but some of their time will become available for other purposes such as research or further course development.

2. In this example the calculations are very sensitive to the cost per hour of teaching. For some elements it may be appropriate to use part-time lecturers or postgraduate teaching assistants at a lower cost per hour. If full-time lecturers were to be used for all teaching the RBL alternative would be relatively more economical than the conventional course, which uses 45% more teaching hours.

3. The cost of delivering course content through lectures in the conventional course is low and any savings in a resource-based alternative can be achieved only be changing the arrangements for seminars and tutorials. It is difficult to deliver material more cheaply than by lectures, though their effectiveness as a means of communication and their impact on subsequent studying may be limited.

4. Print costs as low as those shown here may be possible only with long print runs and a commitment to using the packages for a number of years; all the print costs would then be incurred in year one. A department might have to fund such initial printing, which would exceed normal reprographics budgets. If materials needed to be regularly updated the print costs would rise as print runs shortened.

5. Reprographics and copyright charges are often passed on to students provided that they do not increase the total burden on the students' book purchasing budget. In this example, if students were to be charged approx £15 for the set of six packages the annual saving would be approx £1800, and the saving over five years approx £9000, or 19% of total costs. Recovering production costs, including copyright clearance, is legal under the current copyright arrangements, but making a profit is not.

(cont.)

6. Different institutions pay authors at very different rates for writing RBL materials, ranging from £10 per student learning hour at the bottom end of the market to about £200 per student learning hour at the top end (though even the higher rate is a tiny fraction of the investment in writing time common in the Open University). Some institutions adopt the policy of providing reasonably generous centrally allocated funds for the production of the first set of packages in a subject area, and then use the tutor time saved to produce packages for subsequent course without any further reading.

7. Production costs vary greatly. Some institutions have central units with a team of dtp technicians, word-processing operators, designers, editors, multimedia experts and educational technologists. This kind of provision leads to much higher production standards but very much greater costs than where lecturers do most of the production themselves without additional funding.

8. It would be important to monitor the level of library usage to substantiate the assumption made here that there would be no change in level of use as a result of introducing the RBL course. There could be significant cost implications, and access problems, if usage increased to compensate for the reduction in teaching.

9. More surgery time is allowed for on the RBL course than there was tutorial time in the conventional course. Reliable figures for takeup of tutorial time are hard to obtain, but comparison with usage of surgery time would be important here in order to compare costs. If the demand for surgery time increased as a result of the elimination of lectures and the larger group size for discussion sessions this would clearly limit the overall cost savings of RBL.

Reference:
  Gibbs *et al.* (1994a, pp.28-29)

Activity 18

-   Using the headings in Box 33 calculate the cost of running one of your conventionally-taught modules or units.

-   Think about how it might be presented using RBL as a major component and recalculate the costs.

-   How sensitive are the calculations to the assumptions you have made about the costs of different elements?

-   If you had to design the module/unit within a budget, which over five years would give a 20% cost-saving on your conventionally-taught course, what changes would you make to stay within budget, while minimizing any detrimental effects on the quality of the student learning experience?

6.5   How can RBL be made cost effective?

Before deciding to go down the road of producing a learning package it is important to consider alternative strategies, such as improving the library, developing students' information searching skills, giving a wider range of assessment topics, or arranging for sub-groups to cover topics in rotation. All these strategies will help improve access to limited resources and are generally more cost effective than the 'produce it yourself and give everyone a copy' strategy (Gibbs *et al.*, 1994a, pp.16-17). If none of these strategies meet your needs then there are still several different ways in which the cost of producing RBL packages may be kept low. However, in focusing on 'cost', it is important not to lose sight of 'effectiveness'. The purpose, after all, of all the effort put into developing learning materials is to find combinations of methods which help students learn effectively, albeit within overall budget limitations.

6.5.1   Use an existing learning package

The first strategy involves not producing a package yourself, but using an existing one (Box 34). This is the model which the NCIHE (1997) suggested needs to dominate HE in the future, if numbers of students are going to grow without major increases in expenditure (see Section 6.3).

Box 34:   Buying-in high quality open learning materials from other institutions can be a cost effective strategy, as long as they are used appropriately

The University of Luton, which previously had little large scale experience of distance learning, adopted a policy in the early 1990s of buying-in open learning materials from other institutions, most notably the Open University. "The purchased open learning materials have been used as major and essential parts of the undergraduate provision, with their use being interspersed within and besides conventional teaching. …Off-the-shelf materials have been adopted from around the world and these have been subjected to the University's standard course approval and validation procedures. The Luton approach has provided a greater level of face-to-face support and back-up than would normally be the case for students learning at a distance. …Increasingly, the mixed-mode approach is evolving to incorporate elements of open learning within modules rather than being the chosen mode for the module as a whole. The smaller elements can be either bought in or produced in-house to deal with particular aspects of the curriculum" (Fellows, 1997 pp.149-151). A similar purchasing policy is being applied to computer-based learning materials. A selection of the materials produced from the government-funded TLTP is being incorporated into the Luton curriculum. For example, extensive use of the *GeographyCal* materials is being made in the geography-related courses (see Box 39).

Reference:
    Fellows (1994; 1997)

Not all packages have to be purchased. Informal contacts with colleagues from other institutions can lead to the exchange of resources and information. Some packages have been developed by government-funded consortia and are available free, or at the cost of distribution, to HEIs funded by the government. This applies, for example, to the Teaching and Learning Technology Programme (TLTP) and FDTL programmes funded by HEFCE (Boxes 32, 36 and 39). Some course material is also freely available on the WWW (Box 40). The best example of this in geography is the Virtual Geography Department, based at the University of Texas at Austin, to which several departments in the United States have contributed units and courses to a common format (Box 35).

Box 35:   An increasing amount of teaching material is being put on the WWW

The Virtual Geography Department Project at the University of Texas at Austin, funded by the National Science Foundation, was established to act as a clearinghouse for high quality curriculum materials and laboratory modules that can be used by students and faculty (staff) all over the world. Summer workshops have provided an opportunity for over a hundred geography faculty to discuss, plan and develop materials in many subfields. Working groups have so far been established in ten areas: Cartography; Introductory Human Geography; Cultural Geography; Physical Geography; Earth's Environment and Society; Urban and Economic Geography; Geographic Information Science, Remote Sensing, and Spatial Statistics; Virtual Fieldtrips; History and Philosophy of Geography; and World Regional Geography and Area Studies. Each working group is being sponsored by a different department with members sharing plans and materials under the guidance of a volunteer leader. Stress is being placed on curriculum integration through the creation of on-line syllabi, texts, laboratory exercises, field activities, and resource materials that are of service to instructors in these working group areas. The project links existing materials already available on the Internet, but, more importantly, attempts to develop plans at the subdisciplinary level for the creation of new Web-based materials. To maintain the project after the end of the project funding a Virtual Geography Speciality Group is being established within the AAG.

Further information:
   Ken Foote, Department of Geography, University of Texas at Austin
   email: k.foote@mail.utexas.edu
   http://www.utexas.edu/depts/grg/virtdept/contents.html

References:
   Foote (1998); see also Case Study 4.4 in Jenkins (1998)

6.5.2   Adapt or modify existing resources

Customizing a package to your own needs is an important way of taking 'ownership' of the resources you use. The 'not-invented here' syndrome is a significant limitation to the widespread acceptance of externally produced RBL materials. However, many packages, such as the *Hands On!* modules (Box 11), contain a wealth of ideas which you can modify and adapt for local consumption.

Adapting or modifying them for local conditions can take place at a number of levels. For example, to make the Open University D215 TV programmes (Boxes 24 and 45) more acceptable in North America, two of the programmes, which were thought to have little interest for a North American audience, have been replaced and the remainder have been narrated by an American professional. This 'telecourse' also had an extensive study guide specially produced, which includes extracts from the five UK textbooks, along with learning activities, multiple choice questions, short essay questions, and self assessment questions (Bell, 1996). Each chapter also refers to material in James M Rubenstein's best selling textbook *The Cultural Landscape: An introduction to human geography.* In addition a faculty (staff) guide is available.

Changing the examples or data sets by the tutor can be another way of customizing learning packages. The ability to customize may be important psychologically in obtaining acceptance of an externally produced pack, even if the facility is not used (Box 36). This feature was mentioned frequently as important when the *GeographyCal* modules were being developed, but no one, at the time of writing, had yet taken up the option (Box 39). Designing an assessment exercise around a package is another way of tailoring it to your own needs.

Box 36: **The ability to customize a learning package is important in obtaining its acceptance outside the institution in which it is developed**

Another of the FDTL projects, the Geography for the New Undergraduate (GNU) at Liverpool Hope University College, is developing 24 first-year seminars to develop key skills within a geographical context:

1-3	Establishing a working group: Introduction to geography
4-7	Student skills e.g. essay writing
8-12	Thinking and critical skills: Geographical inequalities
13A-17A	Problem solving and presentation skills A: Physical processes
13B-17B	Problem solving and presentation skills B: Perceptual geography
18-20	Individual oral presentations
21-24	Reflection and evaluation

An important feature of the seminar materials is the ease with which they can be *adapted for different subject content, or for use in a different context.* Their flexibility is demonstrated by the following uses:

- in support of subject lectures;
- as a free-standing course within a module;
- as a study skills module;
- small group work within large groups;
- adaptation for an environmental science module.

The seminars are available in disk or CD-ROM format.

Further information:
Sarah Maguire, GNU Project Manager, Department of Environmental and Biological Studies, Liverpool Hope University College
email: maguirc@livhope.ac.uk

Reference:
http://www.livhope.ac.uk/livhope/gnu/

A variant on the use of existing resources is to present existing lecture material in a different way. This may be provided to supplement lectures, for example, to enable students to revisit the material covered, particularly overheads (for example, Box 25), or it may be provided in printed, audio or video form to replace the lecture (for example, Box 15 and case study in Section 9.1).

6.5.3   Prepare a study guide for an existing text

Study guides are frequently designed to help students use textbooks effectively. They usually contain a mixture of key points, commentaries on issues raised, additional material to fill gaps, elaboration of difficult sections, questions and exercises. They are most commonly used with large introductory textbooks, some of which appear too dry or too daunting to generate much productive student effort, and are more frequently used in North American universities than in Britain. Some study guides are produced by publishers and are marketed with the textbook as a package for use as the course text, others may be produced in-house (Box 37; Section 6.5.2). Generally, study guides are much cheaper to produce than it is to write comprehensive open learning materials from scratch. Even the Open University produces such study guides for their advanced specialist courses, when it is uneconomic for them to produce their own course units.

Box 37:   A well written study guide can assist students to learn productively from a textbook

Haggett's (1983) *Geography: A modern synthesis* was one of the best selling textbooks in human geography in the 1980s. Many universities adopted Haggett (1983) as a course text. For example, Penn State University used the textbook for their introductory course in Human Geography, which was delivered as an independent learning module. They produced their own extensive study guide (253 pp.), more than a third the size of the textbook (Sheridan, 1987). Each of the 19 units in the guide includes the following sections:

- purpose of the lesson;
- instructional objectives;
- reading assignment;
- commentary;
- lesson assignment.

The commentary section includes additional materials, key terms, and self-help exercises. The students who took the course were allocated to an instructor who marked their assignments. Students were encouraged to seek advice and assistance by writing and telephoning their instructor.

References:
   Haggett (1993)

6.5.4   Concentrate on the provision of learning exercises rather than content

An alternative to producing a comprehensive open learning package, or a study guide for a text book, is to devise suitable learning activities. These can generate learning hours much more cost effectively than writing out course content (Box 38). A textbook may be used to provide much of the content of the course enabling more of the contact time between staff and students to be focused on discussions of the material and application of the ideas (for example, see case study in Section 9.4).

Box 38:   Designing an effective learning exercise takes much less time than preparing the course content to go in a learning package

At the University of Plymouth the final year Geohazards module is designed around a series of independent self-learning packages. The module draws students from geography, geology, earth science, environmental science and combined science. Students work in mixed groups of five or six and study one example of a geohazard in great depth. However, they are expected to acquire a good knowledge of other hazards through the Geohazards Conference which takes place at the end of the module. There are no formal lectures in the module.

A number of staff from the Departments of Geographical Sciences and Geological Sciences contribute to the module by offering to advise on a case study of a particular geohazard (e.g. flood, landslide, volcanic, earthquake, coastal, and mining hazards). *A resource-based learning package is associated with each case study*. These introduce the hazard and the case study, define the task, and list the references, maps, aerial photographs and other resources available in a three or four page handout. For example, the students have to produce a consultancy report for the Landslide Hazard Assessment:

> *"The Department of the Environment are preparing a White Paper on the use of landslide hazard analysis to identify landslide prone areas in order that such information can be incorporated in Local Plans. As a specialist consultancy group in geohazard appraisal you have been appointed to undertake an analysis of the various methods that are available, to recommend the most suitable approaches to use and to justify your choices by presenting a worked example of the use of the methods on an area between Seaton and Lyme Regis."*

The module is assessed by coursework (60%) and an examination (40%). The coursework consists of a number of elements which come together at the Geohazard Conference. This all-day event takes place off-campus and includes guests involved with geohazards. Each team makes an oral presentation; produces a poster on their hazard, which is displayed in the foyer of the conference hall; and a team report (8,000 words) in which each individual's contribution is identified (worth half of the coursework mark). The examination consists of a two-hour unseen paper divided into two sections. The first section is a single compulsory question which demands an answer with reference to more than one geohazard. The second section is free choice and consists of one question from each of the staff case-study leaders.

(cont.)

Each group receives on average about 2.5 hours of advice from a member of staff in the preparation of the work. Including the introductory session and the conference, each student is thus 'in class' with staff for approximately 11-12 hours, about half the number of 'contact' hours in a conventionally taught non-laboratory module and a third of the hours of a laboratory module. The staff hours involved increase in proportion to the number of groups to be supervised. With a module of 48 students (8 groups) the total 'contact time' for the six staff involved amounts to approximately 60-65 hours. However, the case studies are based on research areas or field sites with which the individual staff are already familiar, thus they do not involve staff in significant additional work. When this is taken into account, along with the lower amount of additional time spent in assessment, the total staff loading is not that much higher than for conventional modules where a single member of staff teaches the whole class together and all work is individually assessed.

The staff and students involved enjoy the intensive, interactive style of teaching associated with this module. In a previous version two case studies were undertaken and two geohazard conferences; this was changed in response to feedback from student and staff allow more in-depth study of one case study and to reduce the time commitment to the module.

This example illustrates how RBL can provide an in-depth learning exercise, while the compulsory examination question encourages some breadth of study as well. The examination also provides the incentive for the students to participate and learn from the other presentations and posters at the conference. The module gives the students valuable experience of collaborative team work.

Further information:
   Dave Croot, Department of Geographical Sciences, and Jim Griffiths, Department of Geological Sciences, University of Plymouth
   email: dcroot@plym.ac.uk; j1griffiths@plym.ac.uk

6.5.5   Focus on low-cost resources

If, after considering all the alternatives, it is still necessary to produce a comprehensive learning package which includes content prepared from scratch, the main way of keeping the cost down is to use low-cost materials. As we saw in Section 6.4 the relative costs of different media vary widely. Print-based materials are generally the cheapest to produce. Minimizing the amount of copyright material included in the package also helps to reduce costs (Appendix II).

6.5.6   Develop materials collaboratively

Another way of keeping costs down is to develop material in collaboration with another institution or institutions. This may involve more than one group working on the same topic or different topics being divided up between members of the consortium. Working collaboratively can help to spread the costs and may enable more expensive technologies to be afforded (Box 35). This may necessitate concentrating the technical development at one site, while the authoring of the academic materials is spread more widely, as was done with the *GeographyCal* materials (Box 39).

Box 39:   Developing RBL materials in consortia helps spread the cost of development and extends the ownership of the final products

GeographyCal comprises a suite of innovative geographical courseware modules for use on the PC. These modules facilitate efficient teaching and learning environments for core topics, concepts and techniques in introductory undergraduate geography courses. The materials have been specified, designed and developed by a consortium of seventy-two UK higher education institutions. They form a library of high quality transportable resources that use data visualization, dynamic representation and regular assessment exercises to enhance teaching and learning in geography and related spatial disciplines. There are six human geography, seven physical geography and five geographical techniques modules.

Further information:
    Geoff Robinson and John Castleford, CTI Centre for Geography, Geology and Meteorology, University of Leicester and Mick Healey, Geography and Environmental Management Research Unit, Cheltenham and Gloucester College of Higher Education; email: cti@le.ac.uk
    http://www.geog.le.ac.uk/cti/Tltp
    See also Section 10 — Guide to references and resources

Reference:
    Healey *et al.*, 1996b, 1998; Robinson *et al.* 1998

Another advantage of working collaboratively is that it inevitably raises important issues at an early stage, including those of standards, common formats, and flexibility in use. Addressing these issues successfully means that the materials produced should be of a higher quality and are more likely to be used by other lecturers, both within the consortium and outside. Many such groups work essentially on a good-will basis, and this may be sufficient where there are only two or three people involved. Larger groups need a clearer management structure and formal agreements on responsibilities, timetables, quality assurance procedures and so on. Work in consortia tends to progress at the rate of the slowest and written contracts can help members take their responsibilities seriously. International consortia for developing geography teaching materials are rare (Healey, 1998b) (Box 40).

Box 40:   International consortia for developing geography curriculum materials can help to overcome the insularity which often pervades national consortia

Perhaps the best example to date of international cooperation in the development of the geography curriculum is the National Centre for Geographic Information and Analysis Core Curriculum project in which 35 GIS educators in the US, Canada and the UK developed a comprehensive set of lecture notes for teaching beginning GIS professionals. An updated web-based version of this course is in preparation, which includes 76 lecture topics and 19 section editors from five different countries.

References:
    Kemp & Goodchild, 1991, 1992
    http://ncgia.ncgia.ucsb.edu:80/giscc/

6.5.7   Produce a package which can be sold outside the institution

An alternative to going for the 'cheap' solution is to produce a product which can be sold commercially. Many questions need to be answered if this seems to be an option including:

- Is there a market for the learning materials?
- What do the potential purchasers want?
- Is the package self-contained?
- How will the package be used?
- Who is going to do the marketing and the selling?
- Where will it be marketed?
- How many sales are needed at what price to cover costs?
- Who owns the product?
- How is the income to be distributed?

With too many packages, the possibility of marketing them externally comes as an afterthought. If external sales are a realistic proposition then the package needs to be designed from the outset for the potential customers, because it may be used in a variety of ways by staff in different institutions, it is important to design the package for maximum flexibility. However, care needs to be taken that the package will still be suitable for use by your own students. A test of the commercial potential is whether a publisher or software company, who produces similar types of package, would be interested in adding it to their list. RBL is a new area for most companies and the market may change rapidly over the next few years. A major expansion in the production and use of RBL packages will be required if the future scenario painted by the Dearing Committee report is to come about (Section 6.3).

Activity 19

- What *alternatives* are there to the 'produce it yourself and give everyone a copy' strategy to developing RBL?
- Do any of these adequately address the issues you face?
- If not, what ways of producing an RBL package would suit your needs and those of your students and are also cost effective?

7 How can RBL packages be designed to promote learning?

The early published examples of the use of RBL in geography in higher education focused on highly structured introductory courses. Several were influenced by the Keller Plan for self-instruction (Keller, 1968) in which the statement of unit objectives and the self-paced testing system are central. For example, Fox & Wilkinson (1977) and Backler (1979) have used the Keller Plan in North American Universities to bring entrants from varying backgrounds up to a common level. Self-paced schemes were also developed for field and laboratory techniques in physical geography (Cho, 1982; Clark & Gregory, 1982; Keene, 1982). Later published studies (Church, 1988; Fox *et al.*, 1987; Healey, 1992; Jones, 1986) and many of the cases discussed in this Guide describe RBL packages covering a wide variety of subdisciplines within geography and different stages of geography degree programmes. Given the range of uses of RBL in geography (see particularly Section 5) a variety of designs may be used for different purposes. It is, however, possible to extract from the extensive literature on the design of RBL packages (Adams, *et al.*, 1980; Freeman & Lewis, 1995; Kember & Murphy, 1994; Lewis & Freeman, 1994; Race, 1992, 1993b; Rowntree, 1990, 1992, 1994a, 1994b, 1997) some general design guidelines.

I will focus in this section on just two topics. I will begin by highlighting some of the key features of RBL packages and some of the choices available to authors when designing such packages, before going on to examine the different ways in which RBL activities may be designed and assessed and feedback given to learners.

7.1 What are the key pointers to designing a quality RBL package?

There is no magic formula for designing a learning package, because the purpose for writing them varies and the way in which they are planned to be used differ. However, there are pointers as to what could constitute quality (Gold, *et al.*, 1991). Five are discussed below. The first two are phrased as questions and it is suggested that you should make your answers explicit to your students; the other three are phrased as advice and are based on characteristics of packages which provide effective learning opportunities.

7.1.1 Who is your audience?

In many cases the answer to this question will seem obvious: 'It is students taking module GE201'; or 'It is students who are planning to undertake a social survey in their dissertation'. However, once produced and put in the library, or published and sold, students and other staff may use your material in a variety of ways. Indeed, it is a good idea to design the package so that another member of staff in your department could use it when, for example, you are on sabbatical or have different responsibilities.

It is useful when planning a package to write down the characteristics of your potential audience in terms of:

- their background knowledge (for example, what assumptions are you making as to the knowledge and skills your readers need to have to understand your package? Are these assumptions justified for all your audience? Do you need to give additional help or advice, such as on working in groups, undertaking an information search, citing references);

- the different reasons they may have for using your package (for example, to answer an assignment you have set, to plan their own project, to fill a gap in their knowledge);

- the different ways in which they may use the package (for example, reading it in the library, on the bus, while using a computer-assisted learning package, in conjunction with a video).

This exercise should encourage you to be explicit about your assumptions and to design the package so that it can be used flexibly.

7.1.2 What are the aims and objectives of your package?

"Well-defined learning outcomes can lie at the heart of an efficient strategy for designing open learning materials"

(Race, 1993b, p.75)

It is important to make it clear to your readers what it is they are supposed to learn from your package. Sometimes it may also be helpful to indicate what they are *not* expected to have learnt as well. For example, 'by the end of this package you should be able to…, but I would not expect you to be able to… at this stage' (Adams *et al.*, 1980, p.7). Whereas aims generally express the purposes of the writer and give broad statements of intent, learning objectives are much sharper and give specific statements about what the learners should be able to do by the time they have completed the section or package.

Learning objectives are best stated near the beginning of the learning package. They are placed here to guide the learner and therefore are best expressed in ways helpful to the reader (Race, 1993b). Thus it is useful to:

- *Address the learner directly*. To write that 'the expected learning outcomes for this module are that students should be able to…' is not as user-friendly as 'by the time you've finished this module you'll be able to…'

- *Avoid using vague words like 'know', 'understand', 'appreciate', and 'aware'*. It is more helpful to say what the learners should be able to do to show that they know, understand, appreciate, or are aware. A useful list of 'doing' verbs is given in Table 6, Section 7.1.3, where they are related to Bloom's learning taxonomy.

- *Try to avoid using jargon.* State the objectives as simply as possible at the beginning of the unit; technical terms and disciplinary jargon introduced in the package can be used at the end when checking that the students have met the learning objectives.

- *Comment on the different objectives.* In a written list of half-a-dozen learning objectives they all seem of equal importance. It is helpful to add some comments, such as 'this is the key objective', or 'you may find this one difficult at first, but it is important that you pull the stops out and spend a little longer getting to grips with this topic, because it underpins much of what follows'.

- *Make the learning objectives match the assessment requirements.* If the students know that they have achieved the learning outcomes, then they should also know that they will be able to undertake the assessment successfully (Box 41).

Box 41:   It is most helpful to learners if you make your learning objectives explicit and express them in a user-friendly style

Learning objectives can be specified at different levels of detail. Here are two examples from units in two Level one modules at Oxford Brookes University written by John Gold and Jon Coafee, and Derek Elsom, respectively. The first provides three general objectives, while the second gives twelve detailed objectives.

2608 Britain's Environment and Society Unit 3 Reconstructing Britain's Inner Cities

After completing this unit you will be able to:

- describe the origins, meaning and implications of the inner city crisis;

- give details of an important case study of inner city regeneration;

- explain the role that values play in our understanding of contemporary urban problems, and the influence of the media in helping shape these values.

2605 Geography, Environment and Society Unit 3 The Greenhouse Effect and Global Warming

At the end of this unit you should be able to:

a.   explain to say, a member of your family what is meant by the greenhouse effect and to state the most important greenhouse gases;

b.   state by how much average global temperature (in °C) is expected to increase from now until the year 2030 and from now until the year 2100, and explain why these seemingly small figures are likely to be highly significant;

c.   explain why the global temperature rise measured since the late-19th century may not be attributed with total certainty to the human-enhanced greenhouse effect;

d.   explain why higher latitudes are expected to experience a greater temperature increase than lower latitudes;

(cont.)

---

e.   give three examples of how regional climate is expected to change as a result of global warming;

f.   state by how much global mean sea level is expected to rise by the year 2030 and 2100, and name three locations from around the world that are at particular risk from rising seas;

g.   give three examples how global warming may affect human activities (e.g. agriculture) in the developed world and three examples in the developing world;

h.   explain the latest international agreement coordinated by the United Nations Environment Programme (UNEP) concerning what is expected of countries in terms of their emissions of greenhouse gases;

i.   cite three ways in which the priorities, needs and perceptions differ between developed countries and developing countries;

j.   cite one action that an industrialized country such as the United Kingdom or the United States is taking and two actions it believes the international community should be taking with regard to the greenhouse effect and global warming;

k.   cite one action that a developing country such as the Brazil, China or India is taking and two actions it believes the international community should be taking with regard to the greenhouse effect and global warming;

l.   list three skills needed to cooperate effectively with a small group of colleagues in preparing and delivering a spoken presentation.

Further information:
John Gold, Jon Coafee and Derek Elsom, Geography Department, Oxford Brookes University; emails: jrgold@brookes.ac.uk; dmelsom@brookes.ac.uk

Specifying the learning outcomes at the beginning of the design process is also useful for the writer. It should help the author to:

- focus on the key issues and avoid going off at a tangent;

- design appropriate learning activities and self-assessment questions (SAQs);

- set suitable assignment questions.

When writing this Guide I started with the learning objectives to help clarify in my own mind what I was trying to achieve. However, after writing this section I went back and reviewed and rephrased the objectives that stated in Section 1.2 so that they reflected what I had actually written in the rest of the Guide and what I had learnt about defining learning objectives.

Some authors suggest starting with the SAQs (Section 7.2.1) and then adding the words 'you should be able to' before them to define the learning objectives (Freeman & Lewis, 1995). This, at least, assures that there is a direct tie up between the assessment and the objectives.

7.1.3 Build the package around learning activities

"Effective RBL has more to do with designing appropriate learning activities than writing down course content"

(Gibbs et al.*, 1994a, p.23)*

The best learning packages are designed to be different from standard textbooks. They are designed not just to present content, but to provide students with the opportunity to interact with the content (Section 3.2.1). Some learning packages are wholly devoted to the learning process and contain no subject matter, only tasks, methods and assignments. Producing a 'wrap-around' guide to an existing textbook or a set of readings can be a very cost effective method (Section 6.5.3). Even where the package is going to contain content it is often best to start by thinking what you want the students to learn, design the self-assessment questions and learning activities to achieve this Section 7.2.1), and only then develop the content (Freeman & Lewis, 1995; Race 1993b). Your learning package should involve your students in activities which directly contribute to them achieving your learning objectives (Section 7.1.2) (Boxes 11, 37 and 42).

Box 42:   Appropriate learning activities are at the heart of RBL packages

In the first unit of the Level I 2605 course on 'Geography, Environment and Society' at Oxford Brookes University the students participate in a Limits to Growth debate. The seminar groups are divided into two subgroups. Each member of the subgroup is then given a learning pack providing information on either the case for or the case against the proposition that:

"The current rate of population growth is leading to serious global pollution, resource depletion and famine. We are reaching the limits of such growth, which must be GLOBALLY checked."

The packs outline the procedures for the debate, provides extracts from books, and gives questions and answers about the extracts. Assessment is based on the strength of a) how well they assimilate and argue their case, b) how well it is presented. Credit is given for groups who use their initiative and find relevant additional material.

Further information:
     David Pepper and Anna Kilmartin, Geography Department, Oxford Brookes University; email: dpepper@brookes.ac.uk

---

A number of textbooks are now being published, particularly in North America, which are based around active learning exercises. For example, the AAG ARGUS project consists of a backbone text, activities, readings, transparency masters and a teacher's guide — 783 pages in total (AAG, 1995). Although designed for use in secondary school classes, many of the activities are used or modified for introductory geography classes in community colleges and universities. Texts emphasizing active learning are also beginning to appear aimed specifically at higher education geography classes. The *Hands On* project is described elsewhere in this Guide (Box 11). An example of a text produced by a commercial publisher is that by Kuby *et al* (1998), which includes seven computerized projects on a CD-ROM; there is also an Instructor's Web Site.

Active learning relates to the 'doing' stage of learning (Table 2, Section 4.1). It involves providing "opportunities for students to talk and listen, read, write, and reflect as they approach course content through problem solving exercises, informal small groups, simulations, case studies, role playing, and other activities — all of which require students to apply what they learn" (Meyers & Jones, 1993, p.xi). These activities are designed for use in-class as part of an active learning strategy (Moser & Hanson, 1996) and most can be used with RBL exercises in this context (Section 5.2). With learning packages, which students are going to use out-of-class, activities have to be designed for use without the teacher present and often without involving other students. A useful set of ideas about incorporating activities into learning packages is presented by Rowntree (1994a, pp.101-108) (Table 6.

Table 6:   *Designing activities to help learners (Source: Rowntree, 1994a, pp.103-104)*

How might learners make their response?

With different sorts of activity, learners might be expected to make their response in a variety of different forms. Learners might be asked to:

- Just think their response (no record)
- Tick boxes in a checklist
- Press keys on a computer keyboard
- Answer a multiple-choice question
- Underline phrases in a text
- Complete a table
- Fill in blanks left in a sentence
- Write a word/phrase/number in a box or in the margin
- Write or key in a sentence or paragraph
- Write out the steps in a calculation
- Add to a graph, chart, diagram, etc.
- Draw a graph, chart, diagram, etc.
- Make a tape recording
- Take photographs
- *Any others that occur to you? (What?)*

In general, the simpler the form of response required, the more likely learners are to tackle the activity

---

What might learners DO?

Activities may make a wide range of demands on learners. They might, for example, be asked to:

- Recall content

- Restate content in their own words

- Apply their new learning to given examples

- Suggest their own examples

- Compare or evaluate new ideas

- Reflect on how their own experience relates

- Take on different roles

- Practise what they have learned

- Examine new materials in the light of what they have learned so far

- Interview or discuss with other people

- Carry out practical work

- *Any others that occur to you? (What?)*

"Learning activities can be generated through the assessment system (for example, preparing for a written assignment), class contact (for example, preparing for a seminar presentation), or independent group work (for example, a team co-operating on a group project). Activities which are unrelated to assessment, class contact or peer pressure are less likely to be undertaken seriously" (Gibbs *et al.*, 1994a, p.23). If learning activities are not integrated into the text they should be presented separately as part of the course guide or the assignment guide. This has the advantage of enabling what students do with the material to be adjusted from time to time without altering the content.

7.1.4   Adopt a user-friendly tone and style

"Always read your work ALOUD. If it sounds pompous, obscure or long-winded, it probably is"

(Rowntree, 1994a, p.144)

In preparing learning packages it is useful to imagine that you are talking to two or three particular students from different backgrounds and with different interests. Alternatively, if you are not familiar with the students who will be studying your materials, make up the characters of some people. Thinking that you are addressing particular individuals is helpful in keeping your language direct and straightforward and your style reader-friendly without being patronizing. Keep asking yourself how they would react to what you are saying. You may find it hard to adopt this style of writing, I certainly have! After all as academics most of us have been trained to write objectively and impersonally. I, for example, am used to deleting 'I' and 'you' from student essays and I have not found it easy to change my style in writing this Guide. You are the judge of how successful I have been. Rowntree (1994a, pp.137-144) again provides some helpful advice on writing user-friendly materials (Table 7).

Table 7:   *Writing user-friendly materials (Source: Rowntree, 1994a pp.139-140)*

Making it reader-friendly

An example from course materials

"The term ideology is used in this module to indicate a system of interpretation or belief held by a particular group or class (a political party, profession of voluntary organization, for example). The frameworks of understanding and interpretation contained within an ideological position are likely to represent and express the material and cultural interests of the particular group in question, and it is this factor which gives the concept of ideology its political dimension. Although a system of beliefs requires some degree of coherence to qualify as an ideology, elements of internal contradiction may also be expected as individuals have to continually reassess and recreate the terms of their understanding of events. Some ideologies therefore can be defined precisely, but some are more diffuse, and it is common for ideological differences to exist between individuals who are apparently members of the same group."

A possibly more reader-friendly version

As you've just seen in those two examples, different groups of people have different ways of looking at the world. This brings us to a term we'll be using often in this module — ideology.

What do we mean by ideology? The term refers to the set of beliefs that are common among a given group of people. Different political parties have their different ideologies. So do different professions. So do people in different organizations. They all see the world differently.

You'll usually find that a group's ideology tells you about what it wants for itself and how it plans to get or keep whatever power it has. This is why ideology is a political concept.

We don't normally say a group's set of beliefs is an ideology unless people in that group agree about most of them. However, this is not to say that all members of a group will agree in every detail. Individuals are always likely to be rethinking their positions on this aspect or that. So, although we can define some ideologies precisely, others may contain a range of views. That is, people in a group may agree on most issues but, on some, individuals may show what we call 'ideological differences'. Let's look at some further examples.

How to be reader-friendly

Be conversational — plain-speaking — welcoming

To be conversational:

- Refer to yourself as 'I' (or 'we' if appropriate) and your learner as 'you'.

- Use contractions (you'll, that's, could've, etc.) wherever it sounds natural.

- Why not use rhetorical questions (like this one)? They can help keep learners awake.

- Keep on your learners wavelength — with references and analogies that touch on shared everyday experience.

- Exploit the 'human angle' — by relating your subject to people wherever relevant.

To be welcoming:

- Say who you (and any co-authors) are and tell of your own experience of the subject.

- Remember that your learners may be of more than one sex, race, religion, age-range, level of physical ability, sexual orientation, and so on.

- Avoid language and examples that exclude or offend any of your learners.

To speak plainly:

- Cut out surplus words — e.g. not 'at the present moment in time' but 'now'; not 'aggravation of the redundancy situation' but 'more sackings'.

- Use short (and usually more familiar) words — e.g. not 'utilization' but 'use'; not 'dissemination of misinformation' but 'spreading lies'.

- Introduce specialist words ('jargon') with care.

- Prefer concrete to abstract words — e.g. not 'mortalities were occasioned' but 'people were killed'.

- Prefer active rather than passive verb forms — e.g. not 'changes were made' but 'we made changes'.

- Keep your sentences short — i.e. rarely more than 20 words — but vary their length.

- Keep their structure simple. (Watch those sub-clauses!)

- Keep paragraphs short (e.g. no fewer than three of four per page).

- Use helpful headings.

7.1.5   Avoid overloading your students

It is easy to make unrealistic expectations about the amount of material which your learners have time to work through. Rowntree (1997) provides estimates for the time taken to read text material for study purposes. He suggests an average student might be expected to read 100 words a minute for an easy text (e.g. narrative style) and 40 words a minute for a difficult text (e.g. complex argument). However, some time — say a third — will be needed for your students to work through the learning activities. This would mean that with five hours of study time your students might reasonably be expected to read between 27 and 67 pages (at 300 words per page) depending on the level of difficulty. If a larger proportion of time is spent on activities, then learners will have time to read fewer pages. This workload will include articles and pages from text books as well as material you have written in the learning package.

Activity 20

Take a learning package which is of interest to you, perhaps one which you use or have written yourself, and evaluate it against the points made in each of the sections 7.1.1-7.1.5. Think of ways in which the package could be improved.

Evaluation	Possible improvement				
Is it suitable for the audience?	1	2	3	4	5
Are the aims and objectives clear?	1	2	3	4	5
Are the learning activities suitable?	1	2	3	4	5
Is the tone and style user-friendly?	1	2	3	4	5
Is the workload appropriate?	1	2	3	4	5

Evaluation: 1 = very unsuitable/inappropriate to 5 = very suitable/appropriate

---

7.2   How can RBL activities be designed and assessed and students provided with feedback?

Students are assessment driven. Hence the design of assignments and the assessment system is critical in persuading them to engage with the material and to undertake the learning activities (Bradford & O'Connell, 1998). Assessment needs to focus on testing whether students have met the course objectives and it needs to be undertaken in ways which support learning (Box 43).

Box 43:   Appropriate assessment tasks can transform the way students work with resources

Assessment tasks need to embody some or all of the following features. They should:

- require material from several sources to be collated and interpreted or criticised;

- require the use of concepts or techniques to analyse cases;
- require problem solving and tackling of tasks different from those encountered in the text;
- require interaction between students before they can tackle tasks;
- require creativity and unique application of ideas through the design of tasks.

Source: (Gibbs, 1994a, p.24)

A distinction is often made between formative and summative assessments. *Formative assessment* is designed mainly to help learning take place; while *summative assessment* is designed mainly to check what learning has taken place. Unfortunately in the absence of peer pressure or class contact, which characterizes much RBL undertaken out-of-class, many students do not take assessments, whether formative or summative, seriously unless it 'counts' towards their grade (Section 7.1.3). Hence in designing assessment activities for RBL it is important to ensure that you make them relevant and interesting to your learners, and emphasise the benefits spending time doing the assessments. This may involve persuading the students that they will learn more by undertaking the activities and can check their understanding and progress by attempting the self-assessment questions (Section 7.2.1). This should then enable them to score better in the assessments which count!

Exactly the same criteria of good assessment practice apply to assessing RBL as assessing other forms of learning (Box 44). Most forms of assessment are also applicable to testing what the students have learnt from the RBL materials, whether that is in the form of tutor marked assignments, projects, computer marked multiple choice tests or examination essays. However, one form of assessment which is much more common with RBL than with other forms of learning is *self-assessment*. This involves learners judging for themselves whether they have achieved a piece of learning. You can help your learners do this effectively by devising appropriate SAQs and activities and providing feedback on their responses.

Box 44:   Criteria for good assessment practice

Valid	Frequent	Relevant
Reliable	Criteria are public	Efficient
Fair	Accessible	Informative

Source: Lewis & Freeman (1994, pp.22-23)

7.2.1   Designing self-assessment questions (SAQs) and activities

SAQs and activities are closely related, but differ in their objective. Whereas "activities promote learning, SAQs confirm learning" (Freeman & Lewis, 1995, p.42). Activities are broadly formative, while SAQs are broadly summative (Box 45).

SAQs enable learners to:

- check on their progress;

- identify areas where they have not achieved the learning objective;

- avoid carrying errors or misunderstandings forward.

Identifying what it is you want your students to learn is a useful first stage in devising SAQs. It is likely that you will have a range of learning objectives. So it is useful to check that your SAQs are appropriate to test these. Table 8 presents Bloom's (1956) hierarchy of learning objectives along with examples of key words and question methods which can be used at each level.

Table 8: *Key words and question methods to test different learning objective levels (Source: Freeman & Lewis, 1995, p.51)*

Key Words	Question Methods
Level 1 Knowledge Define, describe, identify, label, list, match, name, outline, recall, recognise, state *(N.B. In HE, knowledge is usually only tested at the comprehension level)*	Multiple-choice True/false Matching Fill in blank Short answer
Level 2 Comprehension Convert, distinguish, estimate, explain[*], extend, generalise, give example, infer[+], interpret[†], paraphrase, predict, rewrite[‡], translate, transform	Multiple-choice Short answer
Level 3 Application Calculate, change, compute, demonstrate, discover, manipulate, measure, modify, operate, predict, prepare, produce, show, solve, use	Multiple-choice Short answer Problem solving Production
Level 4 Analysis Break down, differentiate, distinguish, illustrate, infer[+], outline, point out, select, separate, subdivide	Multiple-choice Matching Mark diagram or chart Short answer
Level 5 Synthesis Categorise, combine, compile, compose, create, design, devise, discuss, organise, rewrite[‡], précis	Short answer Produce item (e.g. plan, brochure, spreadsheet)
Level 6 Evaluation Appraise, compare, conclude, contrast, criticise, discriminate, explain[*], judge, justify, interpret[†], summarise, support	Multiple-choice (with difficulty) Short answer

[*] Occurs at both levels 2 and 6   [†] Occurs at both levels 2 and 6
[+] Occurs at both levels 2 and 4   [‡] Occurs at both levels 2 and 5

Box 45: Activities and SAQs are important methods for helping students to learn and judge whether they are meeting the learning objectives set

Some SAQs can just as easily be used as activities; the difference, as pointed out at the beginning of this section, is in the objective. Activities are designed to encourage learning, while SAQs are there to check whether that learning has taken place.

In the Open University course D215 *The Shape of the World: Explorations in Human Geography* course activities are integrated into the chapters (see also Box 24). Most of them are in the form of questions for students to reflect upon before the author comments on the answers. For example, in chapter 1 on Geographical Imaginations, Doreen Massey presents three newspaper cuttings about a social land use map produced to help Native American Indians in Honduras reassert their rights to exist and to own their own ancestral lands. She then asks the students, in what is the first activity in the course:

"Before you read on, consider what your own reactions were as you read through the cuttings. These will be your immediate reactions, and they may well change or become more complex as you consider things further. Don't worry about that. For the moment just jot down your initial ideas. For instance:

- Whose place is this?

- Does it belong to 'local people', or do people who arrive later have a right to it as well?

If you find these questions difficult to answer, well and good!" (Massey, 1995, p.15)

In the American Study Guide, which uses extracts from this course, questions are included for students to answer having read the text and watched the associated videos (Bell, 1996). For example, the questions on the excerpt from Massey's chapter include:

- Explain the following statement: "No world map is neutral, and none is totally true."

- What is meant by the term *geographical imagination*?

- In what ways is travel writing a means of representing the world?

- What does it mean to describe maps as being socially produced?

- How are the notions of *local* and *global* illustrated through maps?

Whereas Massey uses questions to stimulate learning, Bell uses them for students to assess what they have learnt.

References:
  Massey (1995); Bell (1996)
  see also Box 24 and Section 6.5.2

7.2.2   Providing feedback to SAQs and activities

When teaching in a face-to-face situation, particularly in tutorials, seminars and practicals, we spend a lot of time giving formative feedback through the comments we make, the questions we ask, and even our body language. Much of this is unconscious application of good formative practice. When we design RBL materials we need to give more deliberate consideration to how formative assessment and feedback can be incorporated, either as part of the materials themselves or provided through another mechanism. If using RBL reduces the opportunities for informal formative assessment and feedback we need to build in other formative feedback devices, such as activities, SAQs, and suggestions for group work in informal learner-led groups (Lewis & Freeman, 1994).

Probably the most common way of giving feedback in learning packages is in a question and answer format. The advantage of this technique is that it mimics a conversation in a tutorial class. However, where the commentary immediately follows the questions, many students may simply read on without attempting to answer them. The students may nevertheless still gain from this interrogative style of writing, rather like other students in a tutorial class can gain from listening to someone else answering a question. Questions involving 'choose from the following', 'true/false', 'single word answers', 'put the following in sequence', or 'fill in the gap' are more likely to be attempted than open questions requiring lengthy responses. Incentives to answer the questions can be given by having the students report back their answers in a later class, perhaps as part of a group oral presentation, or by having their answers assessed, perhaps through a computer marked assignment, before providing feedback (Box 46). In some cases questions are asked as prompts for further study and reflection without any formal assessment or feedback being given.

Box 46:   Students need to know how they are progressing when working through independent learning activities

The Environmental Monitoring Technologies module at Middlesex University includes a RBL package. As part of the package students are given access to computer data files which they analyse and produce graphical and statistical output. Because the questions demand semi-discursive answers it is not possible to computerize the feedback using a computer-based test program such as QuestionMark or an authoring language, such as ToolBook. Instead formative feedback is provided by the supervising technician, who is an expert on the subject. He scans each small unit of work undertaken by the students and only issues the next unit when he feels the exercises have been adequately undertaken and the principles behind them properly understood.

Further information:
Ifan Shepherd, Marketing Academic Group, Middlesex University Business School; email: IShepherd@mdx.ac.uk

---

Activity 21

Take a topic that you are thinking of converting into a RBL package and (preferably with a colleague) brainstorm all the ways in which you might build assessment and feedback into the course. When you have exhausted your list, evaluate them against the criteria for good assessment practice in Box 38.

Criteria of good assessment practice	Evaluation of assessment and feedback ideas				
Valid	1	2	3	4	5
Reliable	1	2	3	4	5
Fair	1	2	3	4	5
Frequent	1	2	3	4	5
Criteria are public	1	2	3	4	5
Accessible	1	2	3	4	5
Relevant	1	2	3	4	5
Efficient	1	2	3	4	5
Informative	1	2	3	4	5

Evaluation: 1 = completely fails to meet this criterion to 5 = fully meets this criterion

Based on Lewis & Freeman (1994, pp.20, 22)

8 How can students and staff be supported?

"In the end it is teachers, librarians and students who implement RBL, not policies, management, learning resource centres, open learning units or teaching committees. ...As some of those institutions who have sought most publicity about their radical approaches have found, RBL does not automatically work and is particularly unlikely to work well where there is little support for lecturers and little expertise in delivering RBL courses. ...Achieving quality involves clear policy and strategy, organisation, resources, challenge, support and commitment. There is seldom a shortage of challenge, and there is sometimes no shortage of policy, but clear and comprehensive strategy is less common and adequate support for and commitment from lecturers seem relatively rare"

(Gibbs et al., 1994b, pp.23-24)

Although it is argued in the introduction that students and staff are already familiar with some forms of RBL (e.g. the library, course handouts, practical exercises), other forms of RBL (e.g. the use of learning packages) are likely to be new to many of them. Hence it is important that support is provided to both students and staff, otherwise the potential learning benefits may not be realised. Students need to be introduced to the way in which the learning materials are going to be used and how they fit into the overall course. They need to develop appropriate skills to make the most of this form of learning; realise where they can go for help; and be made aware of the various facilities which are available in the department and institution to help them make a success of the various learning activities which they are asked to undertake. Staff also have need for support. If teaching in a team they need to agree how the RBL materials are going to be used. For staff, support is probably most important in making them aware of the potential and pitfalls of RBL and in facilitating the development and use of RBL materials.

8.1   What support do students using RBL need?

8.1.1   Induction, tutoring and monitoring

"Offering D-I-Y learning through packages without offering learners a decent support system can look very attractive to those who are hoping to pay fewer teachers... to process greater numbers of learners. But paying for less tutorial time can turn out to be a costly economy"

(Rowntree, 1992, p.73)

It was argued earlier (Section 3.3.2) that learning packages are not enough, there is still a need for staff-student contact. Learners without support are more likely to delay the completion of their studies or drop out altogether. Students need at a minimum to be introduced to the learning packages they are going to use, what the objectives are, how they are going to be assessed, and what they should do if they need help. If the amount of learning time a package is designed to support is relatively low (e.g. less than, say, three or four weeks work) then this may be adequate. However, if the whole course, or a major part of it, is taught through learning packages then consideration needs to given to providing tutorial support and monitoring the progress of students. In this context the function of the tutor is not to 'teach' the material, this should have been done by the packages, but rather is to encourage the students, help them interpret and build on what they have learnt, and assist them to understand any of the parts of the package with which they are having difficulty.

Monitoring the progress of students most commonly involves in-course assignments or tests, which, when assessed, give students and staff feedback on how the students are doing (Section 7.2). Occasional compulsory sessions, which provide an opportunity to discuss course topics are also useful, not only for monitoring progress but also for providing the social glue to connect the students to the course and each other. Special sessions for students having difficulties may be another helpful way of providing them with support. Students also need support where RBL is used in other ways, for example, in implementing guides on peer and self-assessment, peer-tutoring and tutorless seminars (Box 47).

Box 47:   When introducing tutorless groups students need training, a clear agenda set for the meetings and feedback on what they are achieving

Tutorless seminars were introduced at the University of Portsmouth in the final-year core module, 'Society, Nature and Place', as a response to increased student numbers and a way of trying to empower students in their learning. Between 125 and 140 students took the module in the first three years of its operation. Seminar groups comprise 12 to 14 students allocated according to their option choices. They are given a half-day training workshop on deep learning and on how to work more effectively in groups. Each group is given a checklist of suggested conventions for group working.

Each week a series of about eight to ten discussion points arising from the weekly lecture is given to each student, with the suggestion that these points form the basis of the group's activity. Some of the points simply underline key terms, while others pose questions about concepts or seminal papers, and often there are suggestions that students find exemplars from their option courses or rework option course material from different theoretical positions.

Each group writes a monthly report to one of the three members of the staff team, who also makes unannounced visits to the group during the first month to check progress. The use of the time is negotiated with the member of the staff team they have approached. Usually staff are asked to take a question and answer session or, more rarely, to debate or to give a mini-seminar paper. Further lectures are explicitly ruled out.

To run 12 weekly seminars with a member of staff continuously present would require 144 staff hours per semester. The structure adopted requires around 55-60 hours per semester, including the training session.

Further information:
    John Bradbeer, Department of Geography, University of Portsmouth
    email: bradbeer@geog.port.ac.uk

Reference:
    Bradbeer (1997)

8.1.2   Developing independent learning skills

The introduction of RBL may involve students coping with different demands from those placed on them in conventionally taught courses. These can generate anxieties (Noble, 1980). Providing appropriate support is therefore an important element of RBL, but students vary in the amount of support they require and should need less as they become familiar with this way of learning. After all the aim of the support is to encourage the students to become as independent as possible. RBL will involve students undertaking more learning on their own, but in many cases it will also involve them working with other students in groups. For students to work independently effectively it is important to ensure that they have developed a range of independent learning skills, including those of finding information, learning effectively from different media, time-management, working in teams, and project management. Providing guides to these skills is a first stage, but just as important is giving students the opportunity to practise and reflect on the skills and giving them explicit feedback on how they can enhance their skills further. Ideas from last year's students on how to succeed with RBL can also be useful (Box 48).

Box 48:   Integrating skills-based workshops into mainline courses can be an effective way for students to learn independent skills

The Issues in Geography module at Coventry University was designed as a 'capstone' core module taken by all final-year geographers. In the first two years it ran with 64 and 75 students.* The course was designed by a team of four staff (a geomorphologist, a hydrologist, a social geographer and an economic geographer) to:

- build on skills that the students had gained earlier in the course, particularly in their placement year;

- encourage students to evaluate critically both their own contribution to group work and that of others and to reflect on the process of working in teams;

- offer a course which would maximize student choice, encourage active learning, afford critical reflection, enable individuals to experience a variety of different tasks dependent on a broad set of transferable and geographical skills, and foster an interest in holistic understanding, particularly with regard to interlinkages between different aspects of geography.

The course team saw their role as facilitators. Teaching and learning was seen to be an interactive process between tutor and student, in which students were encouraged to negotiate their own strategies and outcomes, within defined guidelines. Given this, it was decided that the module would run without lectures, but access to tutors and other students would be an integral part of the teaching and learning process. The course was based around a residential fieldcourse and two main group projects. One project concerned a choice of a local geographical issue, the other involved the role of geographers in understanding a range of environmental hazards.

Skills-based workshops were integrated into the module at various stages. The main ones were involved with developing the skills of working in groups and giving oral presentations. The students were encouraged to monitor their own skills, evaluate the effectiveness of group work and reflect on whether the course has been a value-adding process by completing the following reports:

- *Skills audit* At the outset and on completion of the module, students were required to carry out a personal assessment of their own level of transferable skills (communication skills, interpersonal skills, geographical skills).

- *Group-work log and group-work report* The students were required to keep a log of their group-work activities. This represented not only an inventory of activities carried out, but also a reflection on how well the group was functioning. Also, on completion of the module every student provided a synopsis of what they had obtained from the group work.

- *Group contract* Each group was encouraged to set its own learning contract, so that all members of the group were made fully aware of their responsibilities to fulfilling the group's aims.

- *End-of-module evaluation* On completion of the module each student was asked to complete an anonymous questionnaire and to consider whether the module had been a value-adding experience. A structured group discussion was also used to identify those comments about the course over which there was general agreement.

* The course described here ran for two years in 1992/3 and 1993/4; it was substantially modified when three of the original four staff who devised the course moved to other jobs during 1994

Further information:
Mick Healey, Geography and Environmental Management Research Unit, Cheltenham and Gloucester College of Higher Education
email: mhealey@chelt.ac.uk

Reference:
Healey *et al.* (1996a)

8.1.3   Facilities required to support students using RBL

Where extensive use is made of RBL it changes the demands for space and facilities to support student learning. Space needs to be made available for students to work individually and in groups, often making use of specialist or restricted materials. Facilities of this kind are often provided in central Learning Centres, but several departments have developed their own resource centres. This not only gives students a place in the department in which they can work when not in classes, but also increases their sense of belonging and ownership of the department, particularly if they are involved in running the centre. At the University of Liverpool "the Resource Centre is now so much integrated into our teaching, it would be hard to imagine how we could manage without it. The wonder is why it is not something which has been taken up elsewhere in geography. In Liverpool we have been the model for resource centres in some other departments" (Bill Gould, personal communication 30 May 1998) (Boxes 49 and 50).

**Box 49:   A resource centre is an important facility in supporting
students using RBL**

In response to a growth in student numbers, a move towards more self-
directed study and a recognition of the diverse nature of the student body, the
Department of Environmental Sciences at the University of Hertfordshire has
produced many readers, course files and case-study packs over the last few
years. These are placed on reference in the Natural Sciences Open Learning
Centre which houses a wide range of interdisciplinary and multidisciplinary
catalogued materials enabling the Centre, in part, to function as the
Environmental Sciences 'laboratory'. The Centre, open nine hours a day
during term-time weeks, is staffed by two part-time technicians and placement
students and comprises a suite of rooms with a capacity for about 50 people
including quiet areas and areas for group work where active learning is
encouraged. This ready availability of resources allows students to study at
their own pace and encourages independent learning.

Further information:
   Jennifer Blumhof, Division of Environmental Sciences, University of
   Hertfordshire; email: J.R.Blumhof@herts.ac.uk

**Box 50:   A student-run department Resource Centre has the double
advantage of providing additional learning resources and
giving a team of students the opportunity to develop
management skills**

The Department of Geography at the University of Liverpool established a
Resource Centre in 1991, with a student-led strategy from the beginning. It
was created in the first instance with Enterprise in Higher Education money.
Bids are called for each year in January from student teams of about twelve
(mostly Year 2/3 but with some Year 1/2) to run the Centre from April-April. In
most years there have been at least two rival bids to evaluate by the
Management Committee, chaired by a member of staff and consisting of
another staff member, a representative of the student Geography Society and
two members of the current operational team.

The Centre has boxes for filing materials for each course, with handouts, key
references and other resources. Advice on what could be held in the Centre
without requiring copyright fees was obtained from the University Library,
bearing in mind copyright legislation and the Copyright Licensing Agency
Library Licence regulations. The material in the boxes is additional to that
available in the Library, and includes much ephemeral literature, of the kind that
most libraries do not hold. The main business of the Resource Centre is in the
lending of offprints and other materials.

Income is generated by the photocopying machine in the Centre, as well as the
sale of stationery, discs, and other materials. Students borrow, but also often
photocopy articles, and this generates enough income to pay the student team
@ £3 per hour worked. The Centre is open for 30 hours per week, 10.00-16.00
Monday to Friday, in term time. There is also a surplus for the student team to
do stock-taking and general management outside these hours — for example,
in vacations. It usually works out at c150 hours each month to be paid. The

Department occasionally puts some resources in, but in principle the Centre is designed to be — and largely is — self sustaining.

The Centre has several advantages. Staff can add materials and direct students to them quickly. It provides access to resources that the students would not otherwise easily obtain and takes some pressure off the Library, although there is a danger that the students may over-rely on the materials in the Centre and not use the Library adequately. The student teams also gain management and inter-personal skills. In every year there has been some continuity in the teams from one year to the other.

Further information:
Bill Gould, Department of Geography, University of Liverpool
email: wtsg@liverpool.ac.uk

Activity 22

- What are the strengths and weaknesses of the systems to support students in your department using RBL?

- How good are your students' independent learning skills?

- What can be done to improve the support your students need?

Brainstorm your ideas with a sympathetic colleague and then go and discuss them with your department Chair and/or put an item on the agenda of the next department committee meeting.

8.2   What support do staff need in developing and using RBL?

8.2.1   Institutional and departmental support

Staff, as well as students, need support if RBL is to be implemented successfully (Box 51). Support may be helpful at three main stages:

- raising awareness of the opportunities and pitfalls of developing and using RBL;

- developing RBL materials;

- using RBL with students.

Awareness may be raised through providing staff with material about RBL, such as this Guide. Running workshops at which the ideas presented in these sources can be discussed are also valuable (Box 52). Research indicates that non-users of RBL are particularly concerned with the effects that introducing this innovation will have on them personally. Until these concerns are addressed non-users are unwilling to listen to or accept potentially valuable information about the benefits of RBL to students and how to manage RBL (Baxter, 1990b). Curricular and pedagogic change is an intensely political process and is closely bound up with the working context within which such change is promoted (Jordan & Yeomans, 1991).

Box 51:   Institutional and departmental support is vital for staff implementing RBL to help them change their working methods and develop new concepts of teaching and learning, together with new skills and values

Has your institution and/or department:

- Created a working milieu in which colleagues' differing perspectives and experiences can be reflected upon, freely discussed and allowed to influence the way that RBL develops?

- Made a public statement of its intention and quality assurance standards for the use of RBL?

- Rethought its use of space (e.g. teaching spaces. Learning centres. Library workshops etc.) to facilitate a shift to RBL?

- Encouraged staff involvement by measuring their output in terms of how many hours of learning they support rather than how many hours they teach?

- Made teaching excellence an explicit and high priority criterion for the promotion of staff?

- Released staff from other teaching duties to give them time to rethink their course and develop materials?

- Ensured that staff who develop innovative teaching can expect the rewards to outweigh the hassle?

- Devolved budgets and provided accurate information about costs to groups and project teams so that they can make their own realistic decisions in costing RBL?

- Minimized any obstructions from committees controlling resources or likely to delay decision-making?

- Shifted funding from delivery costs to development costs, and shown itself willing to provide up-front investment?

- Provided printing and other appropriate production facilities enabling staff to produce multiple copies of materials with minimum fuss and expense?

- Set up a record-keeping that can keep track of student progress and trigger staff action where needed?

- Thought out and resourced a strategy for staff development and sharing or experience about RBL?

- Developed a reward system that — whether through extra money, promotion or enriched opportunities — reassures colleagues that their efforts not only feel worthwhile but are also valued by their institutions?

Reference:
Rowntree (1998, pp.49-50); note the term RBL is used above instead of 'materials based learning (MBL)' which is the term used by Rowntree.

Where extensive use is made by a department of RBL materials a single staff development session is insufficient. New tutors, in particular, need to be provided with on-going support and development (Robinson, 1997).

Box 52: **Workshops provide an important opportunity for staff to discuss the issues involved in developing and using RBL**

A workshop has been designed to go with this Guide. It can easily be tailored to meet the needs of individual departments. The workshop consists of four modules:

Module 1 The nature and use of RBL *(50 minutes)*

Understanding the different circumstances in which RBL may be used and evaluating case studies where RBL has been used effectively in geography

Module 2 Benefits, pitfalls and barriers to developing and using RBL *(45 minutes)*

Appreciation of the benefits and problems for staff and students using RBL and realising the potential barriers for individuals and departments wishing to expand their use of RBL and how such barriers can be overcome/minimised

Module 3 Designing and assessing RBL packages *(60 minutes)*

Appreciation of the ways in which RBL packages can be designed and assessed to promote learning

Module 4 Planning the development of a RBL package *(60 minutes)*

Production of a plan to convert a topic you or a colleague currently teach through lectures, tutorials and/or seminars to a RBL package

Further information:
Mick Healey and Phil Gravestock, Geography and Environmental Management Research Unit, Cheltenham and Gloucester College of Higher Education; email: mhealey@chelt.ac.uk; pgstock@chelt.ac.uk;

To facilitate the *production* of RBL materials support from Heads of Department, library staff, computer technicians, print-room staff and clerical staff may all be required. Staff thinking about developing learning packages need to be familiar with agreed institutional practices in such areas as copyright, production standards, distribution and charging. However, the most important support staff need is in providing time for them to produce the learning packages. With RBL time has to be invested up-front before the students use the materials. This can be facilitated by Heads of Department giving staff a time allowance for developing the materials and/or enabling them to use time saved in using the materials for other purposes. For example, at Plymouth University staff were allocated six 'contact' hours on their timetable for each of the six study packs used in a new course (Jones, 1986). There is a major disincentive for staff to develop RBL materials if as a result they are given additional teaching duties, because of the savings in contact hours they have achieved.

Staff may also benefit from an introduction to *using* newly produced or purchased RBL materials. For example, it is particularly important where team teaching takes place that the team has agreed what role the RBL materials are to play, how they fit into the rest of the course, what the objectives of the package(s) are, and what role the staff are to play in facilitating the students use of the package(s) (Healey, 1992).

Activity 23

What support is there for developing and using RBL packages in your institution?

1. What institutional funds can you apply for to support the development of RBL packages? What criteria are used for assessing the bids? When is the next deadline for applications?

2. Who is responsible for obtaining copyright permissions?

3. Is there an institutional house style for RBL packages? Do you have a copy in the Department?

4. Are there institutional procedures for evaluating the quality of RBL packages?

5. Does the Educational Development Unit run workshops on the use and development of RBL? When is the next one?

6. What support and encouragement will your Department Chair give you to use and develop RBL materials?

7. Who in your department/institution has experience of developing RBL packages with whom you could go and discuss your ideas?

8.2.2 The role of learning-support staff

The development of RBL is blurring the traditional boundaries between academic and learning-support staff. In many institutions the development of RBL packages in geography involves a close cooperation between the geographers and the learning-support staff. It is common for the latter to have the major responsibility for the design, production and distribution of the packages. Obtaining copyright permissions is another role often undertaken by staff in the Learning Centre (Box 53; Case Study 9.2; Appendix II).

It is important that Learning Centre staff are involved in discussions about the development and use of RBL at an early stage, so that the way in which students are going to be asked to use the learning resources housed in, or accessed from, the Learning Centre may be planned. It is small wonder that staff working in the library get a little miffed if 100 students suddenly arrive at the library counter asking the same questions and all wanting access to a key reference, the only copy of which is already out on loan.

Learning-support staff have a key role to play in developing students as independent learners through improving their information literacy. Exercises involving searching the library and using the Internet are often devised by librarians. However, unless they are closely related to the discipline being studied by the student and are linked to the needs of the students at the time the exercises are undertaken, the skills learnt may not be transferred to actual research projects (Bainbridge, 1997; Tierney, 1992). This again illustrates the need for close cooperation between geographers and the relevant learning-support staff.

Box 53:   Learning Centre staff play a major role in the development of readers and learning packages at many institutions

At Cheltenham and Gloucester College of Higher Education readers are produced centrally by the Flexible Learning Design and Development Unit. The Unit, in consultation with the Module Tutor and the Learning Centre Professional responsible for geography, obtains copyright clearance and arranges for the articles, along with a list of contents, to be compiled in a booklet with glossy covers designed in a house style.

Most readers contain 10-12 articles or chapters, many of which are either not held elsewhere in the Learning Centre or for which there is only a single copy. A set of ten readers are produced under the College's agreement with the Copyright Licensing Agency; some are put on short loan, while others are placed on the Open Shelves. The copyright and reproduction costs are charged to the Department Learning Centre Resource Fund. Copyright costs vary widely, some publishers do not charge anything, while Carfax currently charges £28 per article. So far twelve geography-related modules have readers associated with them.

Further information:
   Nicky Williams, Francis Close Hall Learning Centre, Cheltenham and Gloucester College of Higher Education; email: nwilliams@chelt.ac.uk

Activity 24

How closely do you work with the learning support staff in your institution?

1.   To whom would you go to discuss an idea to develop a RBL package?

2.   What are their roles in the institution?

3.   What help can you expect from them?

4.   How can you help them provide a better service for you and your department?

5.   How many of them are members of your department course committees?

9   Case studies

In this section I present five case studies which examine a variety of experiences of using RBL in geography. They are presented in more detail than in the boxed examples used in the text so far.

9.1   Replacing lectures by using audiotapes and written materials

Summary: *This case describes the use of audiotapes and session notes so that lectures can be replaced by seminars. The approach has meant that more students have gained a better understanding of the topics covered. The more able students can use the available contact time to look at the subject material in greater depth. Overseas students, for whom English is not the first language, can control the quantity of information they deal with and the speed at which they study. All students can receive information in a variety of ways so that they are more likely to find an approach which best suits their own learning style.*

9.1.1   Background

The concern which prompted the course designer in the Department of Civil and Environmental Engineering, University of Bradford, to explore the potential of RBL is encapsulated in the question 'What is the best use of my time when interacting with students?' Twelve years ago he began to use audiotapes to support his lectures, and has developed this technique so that the audiotapes, together with 'lecture notes', form the basic method of information transfer. This enables all the contact time available to be reserved for seminar and tutorial work in which the topics studied can be discussed, interpreted, applied and extrapolated.

9.1.2   Context

This case study describes a second-year module from a course – BSc in Environmental Management and Technology. The second semester module, 'Water Quality and Treatment', covers a wide range of topics, including chemistry, biology, microbiology, sociology and engineering. The module is timetabled for three hours a week – two lectures and one seminar. At Bradford a module has twelve weeks of teaching. The course typically enrolls about 20 students.

9.1.3   Aims

The primary aim of the course designer is to deepen the understanding of his students by increasing both the number and quality of seminars. To achieve this, audiotapes and supporting written materials are used to replace lectures as the method for transferring information. More time is thereby made available for seminars and tutorials and students are better prepared for deeper debate and broader discussions.

9.1.4   Implementation

Many of the approaches that have been brought together in the teaching of this course have been developed in a number of civil engineering modules. The different elements have therefore been introduced, evaluated and accepted by other teaching staff as being valid and appropriate.

The audiotapes and written materials are produced in-house by the course designer. The tapes are prepared (see Section 9.1.6) and master copies are kept in a learning resource room, where there are facilities for copying up to eleven 90-minute tapes at one go in just 90 seconds. The students provide their own tapes but only one is really needed. The written materials are available to the students in a bound form at the start of the module. They are also available as 'Word' files.

9.1.5   Student induction

The second-year students involved are no strangers to independent and resource-based learning. They have received first-year courses which develop the transferable skills required, such as problem solving, and courses which utilize those skills. However, an introductory workshop is held at the beginning of the module in which the course designer explains how the course runs and what his role is. The students also have the opportunity to discuss their fears and concerns about operating in this more 'student-centred' way. The workshop uses a mixture of small and large group exercises and therefore a flat-floored room with movable furniture is used.

9.1.6   Resources

The module content is contained on eight tapes with associated notes and problem sheets. A further tape helps explain the final piece of coursework. Preparing the tapes takes a little practice and the following points may help anyone considering trying this teaching technique:

- Do not try to give a straight lecture to the tape recorder;

- A C90 tape is sufficient for about two lectures;

- Break up the material into segments of about 15 minutes, using, for example, a question or brief activity and a short piece of music;

- Carefully refer to any visual material in the notes, giving precise page or figure numbers, and explain them in detail.

The reasons for using audiotapes as the basis of the resource are several. Tapes are inexpensive and easy to update. They also provide a medium which is readily accessible to most students in their own rooms. For those students who do not possess a tape playback machine there are several freely available in the learning resource room. Students can replay tapes as often as they wish and this is particularly helpful for students whose first language is not English.

The tapes are accompanied by very detailed notes, which contain:

- explicit objectives ('At the end of the session you should be able to…');

- detailed summary notes, which do not try to mimic the wording used on the tapes;

- numerical examples;

- diagrams, figures and tables;

- questions, which can be emphasised on tape and used in tutorials.

The problem sheets that accompany each session contain two or three questions which can be used as self-assessment problems or can be discussed at tutorials. In another civil engineering module the course designer has produced a series of audiotapes in which he talks through his solutions to the questions distributed on problem sheets. These tapes are provided in the learning resource room as another resource for any student who wishes to use them. However, in the case of the 'Water Quality and Treatment' module the students work through the questions and the course designer looks at the answers in a tutorial (see Section 9.1.9).

9.1.7   Operation

The 12-week module runs with three hours of contact time per week – this would traditionally comprise two one-hour lectures and a class seminar session. The course content is contained in nine audiotapes and accompanying notes and the three teaching hours available are all used for tutorials. The class is divided into three groups of about seven students for the tutorials, with each group attending one tutorial a week. In most other modules attendance at tutorials is voluntary, but in this case attendance is compulsory because the performance of all the students in the tutorials is one way in which the course designer can evaluate the effectiveness of the teaching approach (see Section 9.1.9).

The students study the tapes at their own pace. Approximately half of them work totally independently and the others work together in informal self-help groups. Groups of three or four students can often be seen listening to the tapes together in the learning resource room.

On average the students work through one 90-minute tape per week. A tape is thought to be equivalent to two lectures, shorter in time but more concentrated in terms of the information it contains. Using the notes, the students then answer two or three questions on the session topic. The primary aim of the problem sheets is to enable the students to assess their own ability against the session objectives. However, the course designer also monitors their performance by looking at their solutions to problems during the one-hour small-group session which they attend each week. To this end the students keep a workbook which they bring to each tutorial.

The module is assessed through a summative exercise, which takes place during the last three weeks of the module. The students undertake a 'real' problem, using all the resources they have available to them. This takes approximately thirty hours. Students, using the problem-solving skills they have developed in an earlier module, decide whether they want to work alone or in a group. At the end of the module they hand in a report of their solution for the staff to mark. If a group submits a joint report, the members of the group must also submit a statement saying either that they all want to receive the same mark or an explanation of who did what in the group, together with an indication of what distribution of marks they wish to see.

The results produced from this assessment method have not been significantly different from those gained in traditional written exams. However, there is an unquantifiable impression that the students seem to understand topics rather than regurgitate information.

9.1.8   Costs

The equipment required for rapid tape-to-tape copying costs in the region of £700 for a system that will copy one master to three blanks. Extra units, which add the capability to copy another four tapes, cost around £350 (1993/4 prices). Tapes are purchased by the students. The accompanying notes are provided by the department and all production and photocopying costs are currently met by the department. The learning resource room is open from 9.00 a.m. to 6.00 p.m. It is unstaffed and has a security code lock on the door.

It takes the course designer about two days to put together a session that includes a tape, notes and problem sheet. Updating the materials is relatively straightforward, but it can be time consuming if the modifications are so major that a new tape needs to be prepared. The production costs are front-loaded, but when compared to a traditional course they are comparable in the short term and can save staff time in the long term.

The student workload is equivalent to a traditional course, but students have greater control over how and when they work. Some listen to the tapes in their cars, others choose to work in a more structured, collaborative way in self-help groups.

9.1.9   Evaluation

To evaluate the quality of learning the students answer a question on the general topic of the module (e.g. 'What is "wholesome" water?'), before and after they have taken the module. In this way the course designer can assess the level of academic development that has taken place more accurately than by using only summative assessment. The 'before-and-after' method gives an indication of the value added and acknowledges the very different starting points of the students enrolling on the module. It therefore provides one way in which the progress of all the students can be monitored. The question it answers is 'Does this teaching approach disadvantage any groups?'

A second way in which student progress is monitored is through the tutorial system, in which student workbooks, containing their answers to the problem sheet questions, are checked.

This enables the course designer to gain an insight into the thought processes of the students, so that he can discover what they find difficult and confusing. The session tapes and notes can be modified accordingly.

Departmental course-evaluation questionnaires have shown that the students enjoy this way of teaching and have no general complaints other than that the tapes can occasionally be soporific. Other teaching staff have noted that the students on this module 'ask very awkward questions'. It is certainly clear that the students ask much 'deeper' questions in tutorials than they did when traditionally taught. This makes teaching a far more interesting, stimulating and satisfying experience.

9.1.10 Developments

'Water Supply' is a second-year module in a three-year course. The course designer has built on the students-experience of independent study in the final year in a self-directed problem-based course. In this module, first offered in 1995/6, students are given the title of the module 'Waste Water Treatment' together with the general aims of the course. The students then decide on a week by week basis, what to study, what the outcomes will be, and what evidence they need to accumulate in order to demonstrate that they have met their self-imposed objectives. At the end of the module the course designer negotiates with each student a 'grade' which is returned to the formal Assessment Board. The course designer finds this module a delight to teach, but accepts it would not suit everyone. The students enjoy it. His contact time with the students is low because they have developed the skills to learn by themselves. Most weeks the students simply let him know what they have done and what they are planning to do, just to reassure him!

9.1.11 Conclusion

This module is just one part of a carefully designed course in which skills and knowledge are co-operatively developed. The use of RBL is part of a wider teaching philosophy which gives greater responsibility to the students and aims to make best use of limited contact time with staff. Contact time is not used for information transfer, but rather to stimulate interest and debate and to solve problems jointly.

One key feature of this module is the variety of learning resources made available to the students. The module caters for a wide range of learning styles so that more students can achieve a higher level of understanding. The audiotapes in particular have provided a freedom for individuals to study in a way that suits them – for example, in the car, in a small group with other students, and so on. This assumption is borne out by students who have asked to use the tapes while on their industrial placements.

The approach therefore helps to support the view of the course designer that 'Learning doesn't stop when you leave university, but is part of a continuing professional development.' The learning that most people do after leaving university is student-centred and resource-based; therefore helping students to study in this way is preparing them for a lifetime of learning.

Further information:

Bob Matthew, Department of Civil and Environmental Engineering, University of Bradford; email: r.g.s.matthew@bradford.ac.uk

Reference:

This case study is an updated version of the one presented in Exley & Gibbs (1994).

9.2 Replacing blocks of lectures with RBL packages

Summary: This case study examines the introduction of course packs into a range of modules in response to an increase in student numbers. The packs typically replaced about a quarter of the lecturing time. The RBL materials were compiled and written by module tutors in consultation with an educational developer working in the library. The students using the packs experience a greater variety of learning experiences. There is evidence that the students have obtained higher marks in objective tests on material delivered through the packages compared with material covered through conventional teaching.

9.2.1 Background

In response to a major increase in student numbers in 1990/91 the School of Geography at the University of Kingston embarked on the production of several print-based RBL packages, usually to replace blocks of between one and four lectures in a range of modules and covering discrete subject areas. The development of the packages was helped by a successful bid to a central Academic Development Fund established to pump-prime changes in learning delivery.

9.2.3 Context

In 1991-92 student FTEs in the School were 350 with an SSR of 19.4 rising to 565 and 28.3 respectively in 1994-5. In 1997-8 FTEs were 437 with an SSR of 19.9. Over 20 geography modules, out of about 90 on offer in the School, currently use RBL packages to cover blocks of material. This is the largest number of packages developed by any single School in the University. The modules currently using the RBL packages vary in size from 40 students to 90 students enrolled. The course is modularized and semesterized with 12 weeks of academic work per semester. The formal teaching programme occupies 11 weeks with the remaining week being used for Field Work or as a Study Week. Timetabled contact time per module is a maximum of 5 hours a week. The RBL packages typically have replaced up to about one quarter of the contact time, although some are used primarily to structure out-of-class activities. The packages cover both human and physical geography modules, with an emphasis on introductory modules. There are also some packages used in techniques/skills-based modules.

9.2.4 Aims

The aims were:

- to give students an improved learning experience by increasing the variety of learning methods used in the course, largely through replacing blocks of lectures with RBL materials;

- to reduce the amount of contact time on the modules where this innovation was implemented so that staff were better able to cope with the increased student-staff ratios that the School faced.

9.2.5 Implementation

The RBL materials were compiled and written by the module tutors in consultation with an educational developer working in the library, who happens also to be a geographer. There are three types of package:

- Study Packs — cover course content and up to three readings;

- Study Guides — include three readings put within a learning framework;

- Resource Packs — consist of collections of supporting reading materials; four copies available for loan only.

The educational developer edits all the packages; checks that copyright permissions have been obtained (Kingston University has a Copyright Clearance Officer based in the library) for diagrams and articles reproduced; designs the packages to be user-friendly and sees them through the printing process. A brief guide has been produced on 'Writing study packages for your students'.

The packages are sold or loaned to students through the library. They are priced at a level (between £2 and £5) to recover some of the production and copyright permission costs. It is cheaper to purchase a package than it is to photocopy it. The resultant income is returned to the School. Experience shows that 85-100 per cent of students enrolled on the modules using RBL buy their own copy of the package and the loan copies are not heavily used. As a separate venture a few of the packs have been sold externally on a full cost recovery basis.

9.2.6 Student induction

The concept of RBL is introduced to the students during induction week when every student is given a free folder in which to store the packages as they are purchased. Each of the packages is an integral part of the module and they are introduced by the module tutor who explains why the package is being used; what the students are expected to do; what they should achieve from using the packages and how they will be assessed.

9.2.7 Assessment

Questions are included in the packages for students to judge their own progress, although the answers are not normally given. The RBL material is assessed in the end-of-module examinations along with other topics covered. A related project on the use of the optical mark reader encouraged several module tutors to introduce objective questions as part of the summative assessment, particularly for Level 1 modules. Objective questions are used in some of the physical geography modules while short-answer questions are more common in the human geography modules. Essay questions based on RBL material were more common in examinations during the early years of the project when it was found that, although fewer students attempted such questions, the marks achieved were either the same or higher than those for other questions.

9.2.8 Costs

Funding made available to the project by the University during 1991/2 and 1992/3 paid for a member of the geography staff to be seconded to the project full-time for the first year as well as providing administrative and technical support, travel, and material costs for both years.

9.2.9 Evaluation

- Strong support from the Head of School led to a large number of packages being developed in the first two to three years; production has slowed since and some module leaders are developing computer-assisted learning packages as an alternative to the paper-based ones.

- A comparison of the marks obtained by students answering examination questions covering the material in the RBL packages with those they obtained on questions taught traditionally by lectures and seminars showed that they did as well or better on the RBL materials (see Box 17 and Figure 4).

- Student end-of-module evaluations rarely highlight the RBL method of teaching either positively or negatively.

- In several cases new lecturers have continued to use the packages which were developed by others who previously taught the modules. A few packages are no longer used because the modules they supported are no longer taught.

- Some of the early RBL packages are now due for revision and updating, but this will require some up-front investment of time.

9.2.10 Conclusion

Kingston University has invested in a sizable collection of geography learning packages (Table 9). Their students have gained from a greater variety of learning experiences. The packages have been used mainly to enable the amount of staff-student contact to be reduced so that the increased student numbers can be managed effectively. The evidence available suggests that the quality of learning has remained the same or has improved with RBL. The School now faces the need to invest further to maintain and update the collection.

Further information:

Sue Watts and Bob Gant, Department of Geography, University of Kingston

emails: s.watts@kingston.ac.uk; r.gant@kingston.ac.uk

(see also Boxes 7 and 17)

References:

Gibbs, *et al.* (1994b, pp.46-7); Rolls & Watts (1994)

Table 9:   *List of learning packages at Kingston University*

Geography Study Packages (SP)

SP1:	Biological basis of evolution
SP2:	The Atmosphere (No longer used)
SP3:	The Atmosphere (No longer used)
SP4:	Aeolian processes and landforms
SP5:	Location of industry and economic activity
SP6:	Principles and results of rock weathering
SP7:	Atmosphere
SP8:	Introduction to soil science (including soil practical)
SP9:	Ecosystems
SP10:	Statistical information sources for the study of social and economic change in the UK
SP11:	Inner city decline
SP12:	Glacio-fluvial and glacio-lacustrine processes and landforms
SP13:	Urban fringe agriculture
SP14:	Social surveys and geographical investigation
SP15:	Questionnaire analysis using SPSS
SP16:	Study skills
SP17:	Stress, strain and earth materials
SP18:	Water and health in England and Wales
SP19:	The advanced urban economy

Geography Study Guides (SG)

SG1:	Deforestation of tropical moist forests
SG2:	Marine pollution
SG3:	Climate change
SG4:	Water and health in developing countries
SG5:	Paris and the Ile de France

Geography Resource Packs (RP)

RP1:   Soil conservation

RP2:   Thermoluminesence

RP3:   Water and health in developing countries

RP3:   Welsh identity

RP4:   Celtic geography

RP5:   Ireland

9.3   Redesigning a module using a RBL package focused around a set of assignments

Summary: This case study examines how a set of carefully structured activities can be used to replace a set of lectures by independent study. The reduced amount of contact time is focused on a series of seminars based around ideas and judgement. The assignments are designed to meet the knowledge and skills specified in the learning outcomes.

9.3.1   Background

The course leader of the final year module 'Heritage Conservation in Practice' at Middlesex University decided to redesign the module using resource-based teaching for three main reasons:

- he did not believe that lecturing benefited student learning because it reinforced a culture amongst students that learning was a passive process;

- he was particularly concerned that his students were not 'thinking' as final year honours students and that there was a need to help them develop the intellectual qualities that would mark them out as graduates;

- due to other responsibilities in the Institution he was not able to commit himself to regular weekly sessions.

9.3.2   Aims

In redesigning the course the course leader had two main aims:

- to move students into active learning through a structured engagement with the course material;

- to get them to appreciate the processes of making a judgement and reaching a conclusion and to understand that decisions are reached on the basis of the quality of the argument.

9.3.3   Student workload

The redesigned course is based around 15 hours of classroom based work and 120+ hours of individual work. This replaced a 14 hour programme of lectures and 8 hours of seminars.

The classroom based work became a seminar built around ideas and judgement. Either the course leader or a student takes a predetermined theme and presents ideas and develops task questions that help others in the group to form their own views. Topics include 'Can we apply ideas about individual personality to a place?', 'What value frameworks are there in heritage management?'.

Most of the individual workload is based around a structured set of assignments. Each student undertakes four out of five topic assignments, a project report, an oral presentation and a seen three hour examination paper. The module mark is based on the project report and the examination, the other two elements are formative, but failure to complete them constitutes failure of the course. Each student completes a weekly learning diary.

Suggested readings and videos are given for each of the assignments. A maximum of 800 words is set for each assignment. The five topic tasks set the last time the module ran were:

- Construct a diagram which shows how a range of agencies and bodies might be involved in the decision to demolish buildings in a conservation area in order to construct a shopping centre and car park. You may make any assumptions you wish. Write notes to explain the diagram.

- You have been retained by the Department of the Environment to draft an executive brief which summarizes the evidence for the role conservation can play in area regeneration. You should identify those situations in which it is/is not effective.

- Take any tourist/heritage townscape (this might be part of London if you attempt this topic during term time or a town accessible to you during the vacation) and *from personal observation* collect evidence on the type and degree of impact of tourism upon the built environment and on urban character. You should explain how you assess 'degree of impact' and how you identify 'urban character'. You should also classify the impacts. Present your results in an appropriate form and comment upon them.

- Summarise the importance of the property market for conservation.

- For any case study assess how far conservation principles were able to withstand growth pressures. By reference to other situations discuss how conservation could have been better supported.

9.3.4   Learning outcomes

A set of outcomes are specified for the course and the relationship between the outcomes and the course activities are mapped (Table 10):

At the end of this module you should have achieved the following outcomes:

Knowledge

- built a model of key agencies, their powers and interests;

- understood how conservation can relate to other social and economic objectives;

- have an in-depth understanding of conservation issues in an area through an in-depth individual study.

Skills

1. improved your research skills in particular

   a) identifying and specifying a 'problem'

   b) analysis and classification

   c) gathering qualitative data of a judgemental or attitudinal nature

2. developed your ability to write succinctly

3. improved your presentation skills

4. understood the process by which you confront a research issue.

Table 10: *Relationships between outcomes and activities*

Outcome	Where achieved						
	Assignments					Project	
	1	2	3	4	5	Research	Presentation
Knowledge 1	X						
Knowledge 2		X	X	X			
Knowledge 3						X	
Skills 1a						X	
Skills 1b	X					X	
Skills 1c			X			X	X
Skills 2	X			X			
Skills 3			X				X
Skills 4			X			X	

9.3.5   Evaluation

Responses via the end of module questionnaire were generally poor because this was the last module the students took. However, those that were returned showed that the students valued the experience because they felt they were working as third year students should work and that they were enjoying the challenges. For some, it raised their self-esteem and self-belief. There were no problems of access to resources because with groups of up to about 20 students the course leader could schedule who did what, when. With larger groups there would need to be a different strategy.

Further information:

Peter Newby, Centre for Higher Education Research, Middlesex University
email: P.Newby@mdx.ac.uk

9.4   Designing a course around a textbook

Summary: The use of a course text was stimulated by the wish to continue to give support to students at a time when modularisation and an increase in numbers was making traditional teaching more and more problematic. This solution emphasizes the use of structured tutorials in which the students discuss their reading, while reducing the time for lectures by half.

9.4.1   Background

The move away from a purely conventional course towards one based on required readings from a standard text and other sources was made as part of the natural development and innovation in teaching taking place in the Geography Division of the School of Natural and Environmental Sciences. A particular issue for the staff was the need to develop group-work skills in students, driven partly by the growing numbers in tutorials, but also by the desire to focus on skills for the world of work.

9.4.2   Context

This is a second-year courses in economic geography which has been in operation since 1988. Between 60 and 80 students take the course each year over 24 weeks. It is one of eight modules which Coventry students take in each year of a three- or four-year degree course, which can include a sandwich placement. The course team comprises three people, two of whom wrote the course text. The other is a research student who helps with the tutorials.

9.4.3   Aims

The course is designed to develop students' knowledge of concepts and theories of location and change in economic geography. It also requires students to deploy group-work skills, as the tutorial groups are as large as 25 and rely for their success on students' working effectively in groups.

9.4.4   Implementation

The textbook was written by two members of the course team in their own time (Healey & Ilbery, 1990). It was introduced experimentally in 1990, with a very strong emphasis on tutorials and without any lectures at all in the second half of the course. Many students were clearly unhappy with this arrangement, so the current compromise pattern was adopted for the first time in 1991/92.

9.4.5   Student induction

There is no induction for students because they have already experienced resource-based learning and group work at earlier stages of the course.

9.4.6 Student activities

The students attend both a lecture and a tutorial once every two weeks. This gives them an average of one hour's contact time per week with about four hours of self-study time, during which they are expected to read the required chapter(s) in the text and study other recommended readings. They also have to find time to complete an essay and a project for assessment. There is a conventional three-hour exam at the end of the module.

Because the students have access to the course text, the lectures can cover more ground than in a conventional course. The staff use the lectures to cover potential problem areas and to discuss additional examples of ideas and concepts introduced in the text. The tutorials use various devices to help the students to reflect on what they have read and heard in the lectures. Thus they might be asked to work in groups to write an essay plan or to compose overhead projector transparencies to summarise the main points of what they have studied.

9.4.7 Resources

The textbook written by the tutors is the main resource, and all students are expected to buy it. In addition they receive a brief study guide which lays out the plan of operation for the course, including dates and times, and set and recommended readings. This is a relatively lean course in terms of staff time required (Table 11).

Table 11: *Staff time involved in delivering the course before and after using RBL*

Before RBL	With RBL
50 students	100 students
Lectures, practicals and tutorials: 3 times 24 weeks	Lectures: 12 hours Tutorials: 48 hours
Marking essays: 50 times 0.5 (25 hours)	Marking essays: 100 times 0.5 (50 hours)
Marking examinations: 50 times 0.66 (33 hours)	Marking examination: 100 times 0.66 (66 hours)
Total 130 hours	**Total 176 hours**
Hours per student = 2.6	*Hours per student = 1.76*

9.4.8 Evaluation

Students clearly like to have a course text that covers the majority of course material. 63% of 88 respondents agree strongly that 'the course text is a great help' (Healey & Ilbery, 1993). There is also some evidence that using a course text encourages students to read more widely — over 40 per cent of those who respond to a questionnaire said they had read at least ten articles or chapters in addition to the course text. However, students' recent coursework suggests that a tightly structured course such as this tends to concentrate academic performance in the middle band, so that there are fewer excellent performances and fewer failures than might be expected on other, conventional, courses.

9.4.9   Developments

Having developed the course the original staff have passed the responsibility for operating it to colleagues. Since then one of the originators has moved to another institution and new topics have been introduced not covered in the textbook. The staff now teaching the course have increased the number of lectures. However, the emphasis remains on tutorial discussions in which the textbook is the core text.

9.4.10 Conclusion

It is not uncommon to have a course text, but in this instance the course team wrote their own, and made sure that there is plenty of support in the tutorials. The case demonstrates that, given sufficient support and motivation, conventional undergraduates can successfully complete a course where contact has been reduced.

Further information:

> Mick Healey, Geography and Environmental Management Research Unit, Cheltenham and Gloucester College of Higher Education and Brian Ilbery, Geography Unit, School of Natural and Environmental Science, Coventry University; email: mhealey@chelt.ac.uk

References:

> This case study is an updated version of the one presented in Cox & Gibbs (1994); see also Healey & Ilbery (1993).

9.5   Developing a set of RBL packages

Summary: This case study examines the philosophy and management strategies underpinning the integrated development of a broad set of RBL materials across a department's teaching portfolio in the early 1990s. The subsequent six years have offered the possibility of evaluating their effectiveness over a more prolonged period than is normally the case.

9.5.1   Background

In 1992, Cheltenham and Gloucester College of Higher Education made a decision to invest £50,000 in a pilot project to develop learning materials to increase the flexibility and effectiveness of undergraduate studies, and to promote independent learning skills. At the time of the development, the College was expanding its student numbers significantly, was part-way through changing its undergraduate courses to modular semesterised patterns, was recruiting increasing numbers of mature and part-time students, and was experiencing the first of the HEFCE Teaching Quality Assessments. After a competitive bidding process, the Department of Geography and Geology was selected to run the twelve-month project, subsequently known as the 'Capitalisation Project'. Following the preparation of academic and business plans in late 1991 and early 1992, the project developed and began to employ teaching materials over the academic year 1992-3. Much of the material is still regularly used and evaluated by students five years later, and over 800 hundred students have engaged with the materials.

9.5.2   Rationale

The main aims of the 'capitalisation' project were:

- to enrich and diversify students' learning in geographical and related areas by providing increased amounts of material for structured independent study;

- to increase the 'student-centredness' of teaching and learning experiences;

- to release academic staff time for research and scholarly activity;

- to demonstrate the potential of new styles of teaching and learning for increasing the flexibility of student attendance at College;

- to respond to the increasing diversity of student backgrounds, attendance patterns and ability ranges, as the scale of recruitment into courses managed by the Department increased;

- to act as a demonstration project for the College, allowing investigation of the most effective method of producing new materials, and informing the senior management of the possibilities and pitfalls of moving towards these approaches to learning.

9.5.3   Context

The early 'conceptual' phase of the project was concerned with establishing the development strategies and educational methodologies to be used, an outline of the academic areas to be addressed, and the administrative arrangements for their production and delivery. There were discussions on the appropriate range and focus of areas to be 'capitalised', the potential benefits of employing and deploying staff at different levels of expertise, the use of academic staff remission, and the role of central College facilities such as the Design Unit. In outline, it was agreed that investment would range across the work of the Department, be focused at module level, utilise the previous developments of experienced teaching staff, incorporate diverse styles of delivery (e.g. text, audio, video, group work), and devolve the presentational elements onto temporary 'junior' staff.

At this stage there was some disagreement about the best way of organising the project with some senior staff within the Department and College, suggesting alternative strategies for the project. For example, among the alternatives mooted were:

- the buying out of full-time academic staff time to prepare materials and their replacement by teaching assistants;

- appointment of a project manager rather than management by an existing academic member of staff;

- using the skills of the College's Design Unit for production of the RBL materials.

The last strategy would potentially allow external publication of any materials developed by the College, albeit with a smaller number of tangible academic outcomes at the end of the process. The Department resisted this approach, wishing to maximise the number of resource-based elements developed, and to explore as wide a variety of approaches as possible, whilst recognising that the quality of final presentation might be lower than would be required for external publication. Subsequent analysis suggests that the central unit would have produced about half the number of modules for the same investment. Although these would have been of a higher quality the potential external market for materials designed to support existing modules was very limited.

For reasons of academic staff engagement, the project managers wished to keep the development as close as possible to the point of delivery, rather than centralising into the College's systems. Within the Department, some insecurity about the longer-term employment implications of RBL was evident, and a level of apprehension about the prospect of changing established successful teaching practices in favour of a more experimental approach. Additionally, there was reluctance to remove academic staff from mainline undergraduate teaching during the project, for fear of compromising the quality of the current students' educational experience. Despite the debate, the project went ahead following the Departmentally proposed strategy.

Once the project had been agreed in outline and funding secured, a member of the Department's academic staff was identified as the project manager, with about 40 per cent of her time allocated to the project. Two assistants were recruited, both enthusiastic new graduates with wide interests in geography and related areas, and with some prior experience of desktop and other forms of publishing. These three staff worked with academic colleagues across the Department to identify a variety of types of material to be 'capitalised', and suggested appropriate styles of delivery. Following a day-conference of all Departmental staff, academic, administrative and technical, firm decisions were taken about the modules to be addressed, from the hard science of geochemistry and global geo-tectonics, though a range of introductory, laboratory and field skills in physical geography, and across to the humanities and social science aspects of human geography. Twenty-one modules were identified, in whole or in part, for 'capitalisation', engaging with all four of the Department's undergraduate fields in Human and Physical Geography, Geology, and Earth's Resources. The emphasis was on the early levels of the undergraduate programme (I and II), and on those areas of the curriculum where the pace of philosophical change was moderate; anticipated 'shelf-life' was an important criterion. The Department's philosophy of Level III tuition being underpinned by cutting-edge staff research suggested this would be inappropriate for 'capitalisation'. A further important criterion in the selection of modules was the personal enthusiasm of the staff member teaching the module for a different, more devolved and participative approach to student learning. The personal benefits for their active participation included the opportunity to enhance significantly the learning materials available to their students; the 'costs' to the staff member included the initial investment of time to reconceptualise the delivery method, and the psychological cost of releasing their teaching materials to more open scrutiny by colleagues than was then the norm.

9.5.4   Implementation

The variety of new developments planned were an orchestrated response to the intentions of module tutors and included new workbooks, readers and resource packs (containing text and other materials) for private study, group exercises, video packages (both in-house productions and new teaching materials centred around commercially available videos), exercises using CAL, and a simulation game. Each module was discussed with the module tutor, clear outcomes agreed and a timetable for activity set.

A wide variety of criteria underpinned exactly how to 'capitalise' each module, including discussion of where control would rest in setting the pace and outcome of independent study, the assessment arrangements, the 'incentives' for students, the style of presentation and its packaging, and the resultant balance between classroom contact and independent (though supported) study. Alternative modes of delivery were explored, including paper-based text, audio and videotape production and in-house CAL. At this stage a very clear decision was taken not to invest significantly in in-house CAL because of the excessively high investment of preparation time required for the limited number of students who would subsequently be able to benefit; within one institution, albeit a relatively large Department, this was not felt to be cost-effective. In addition, the national TLTP project was already developing, and some colleagues were contributing to that. Reinvention was not an option. Similarly it was agreed that the standard of presentation of 'capitalised' material would not necessarily support external sales. The intention was to produce attractive, but not necessarily publishable quality materials, and leave updating possibilities with the module tutors for subsequent years. Some staff visited colleagues in other HEIs to inspect materials in use, to test disciplinary-specific ideas, and to discuss alternatives.

The phasing of developments was designed to allow simpler projects to be tackled first, including those where the module tutor had already piloted some materials. A varied 'diet' ensured that the graduate assistants' learning curve was steep, with their desktop publishing and organisational skills developing quickly and in a complementary way. Skills in the development of audio and videotape were developed later. Each development began with the module tutor, project manager and a graduate assistant examining any existing materials, and discussing the appropriateness of different potential styles of delivery. An outline costing (excluding the time investment) was then prepared for the production of new materials, and a timetable agreed. The graduate assistants then worked sequentially on different modules, each completing about ten elements during the twelve-month period. Their self-reliance and drive were important in allowing the project to succeed.

Detailed discussion of the types of materials developed are beyond the scope of this summary, but the following examples illustrate the diversity of projects which were implemented.

A workbook and other materials were produced for a Level II module on 'River Basins'. This incorporated editing and presentation of three pre-existing commercially or privately-available videos exploring particular geomorphological themes, with supporting papers taken from the research literature and prompt questions for seminars. A short computer simulation on hillslope hydrology (in the public domain) was used as a foundation for exploring in small groups (through structured questions) the movement of water through saturated and unsaturated domains. Documentation to support ten to fifteen hours of structured laboratory work on sediment transport (in a small recirculating flume), sediment rating curves, properties of sedimentary materials (particle size, materials strength), fluid properties (meander simulations on Perspex sheets), and experimental design was prepared. Finally, two field trip destinations (one on mass movement, and one on river channel processes) were set out on worksheets for small groups to use largely independently. This enable one tutor with the help of a postgraduate student to oversee a group of about 60 students. This module was to be supported in its delivery by a weekly lecture programme, and timetabled laboratory access with tutor support.

An option module on 'Fossils and Microfossils' was developed to be suitable for home study using packaged materials. Seven study packs, including boxed rock and fossil specimens, were duplicated using pre-existing departmental resources. A workbook was developed for the module, incorporating a very extensive set of diagrams, and plenty of self-test opportunities. A computer-based classification exercise was produced to underpin the elements of the course dealing with evolutionary theory, and the assessment strategy was refined to take account of the need for students pacing themselves through the module. The reduction in tutor time subsequently required to run this module enabled it to continue to be offered even in academic years when student numbers were relatively small.

'Images of the Third World', a Human Geography Level II option, was partially-'capitalised' through the development of a resource pack dealing with Mexican Art, practical sessions utilising a wide range of advertising posters, and structured seminar material for use with or without the tutor present. A simulation game, concerning the opportunities of individual inhabitants of Third World countries to break out of the cycle of deprivation, was produced for use by small groups of students. A further range of film and video material was purchased, and worksheets developed to underpin student private study of these items. This material was to prove exceptionally popular with students, and the module attracted over ninety students in subsequent runs.

A pre-existing RBL module on 'Aerial Photography and Remote Sensing' was improved essentially through repackaging and enhancement of the items available for home study. Boxed sets of stereoscopic imagery, stereoscopes, hard copy of satellite imagery and short exercises were prepared. These were supported by audiotaped lectures designed to be heard whilst looking at diagrams and photography. The sets included two on aerial photograph interpretation, and four on satellite image interpretation. Early sets explored the principles, and later ones particular applications of the technology. A workbook containing exercises based on the ERDAS image processing system, and a set of local imagery, was also developed to complete the tuition. Typically, remote sensing tuition requires complex organisational arrangements to ensure good access to facilities for students, and in later years, the module was able to be delivered by a range of tutors with less knowledge of the resources available at Cheltenham. The organisational arrangements were particularly popular with mature and part-time students, who enjoyed the opportunity to listen to materials while driving.

A final example is the development of two cross-disciplinary study packs for a module on 'Deserts Past and Present'. The first concerned the relationship between wind characteristics and dune formation, and included satellite image mosaics at various scales, atlas presentations of specific areas, meteorological information and supporting text. Student exercises were constructed to underpin the use of the materials, which were based on two geographic areas where the academic staff concerned had research experience. The second pack explored desert sediments represented in the geological record in the West Midlands, in their local and broader context, and included geological maps, diagrams, photography and text. These packs were designed to reduce the amount of tutor contact required to deliver core material, allowing class time to be spent on fieldwork.

9.5.5   Costs

The development of RBL materials incurs both direct and indirect costs, to degrees determined by the precise nature of the developments. In this case, the total direct cost, as already noted, was £50,000 covering RBL materials in twenty-one modules. Most of the expenditure was on the salaried junior staff and items, which in subsequent years became standard facilities, such as personal computers (two), DTP and authoring facilities, a scanner, video presenters and so forth. Reprographics costs were also substantial in the first year, making up about 10 per cent of the total.

The costs of academic staff time are not included in this analysis (on average about 4 hours per hour of study-time 'capitalised', but a detailed analysis undertaken for two specimen modules (both large Level I modules with student numbers at about 150 per year, and a workbook plus lecture format) suggest that financial savings of the order of 20 per cent of the initial running costs of the module were secured even in the first year of operation, through the reduction in contact time required subsequently for supporting student study. This analysis excludes the base IT costs for the authoring, but includes development time of both junior and more senior staff. In later years, the saving would be commensurately higher, dependent upon the extent of any updating required.

However, it is important to recognise the externalisation of some costs. In later years, some of the ongoing costs were recovered by sales of workbooks to students (reprographics costs only). Students were directed to study off-campus or beyond the departmental boundary in many cases, thus throwing some costs onto central services or the students themselves. The administrative costs were in some cases higher, academic staff time being replaced by administration of study pack loans, self-timetabled laboratory work, or room bookings for student groups. In practice it is difficult to disentangle costs of the new arrangements from increases in running costs which would anyway have been incurred through increased student numbers. Costs per capita suggest a two or three year 'half-life' needs to be anticipated for the content of study packs and work books, after which updating will be required. In those areas where the subject material changes fundamentally very quickly (for example, in areas engaging with legislation, or cutting-edge science), investment in resource-based approaches needs to be more cautious.

9.5.6   Evaluation

For many academic staff the effort they invested in this programme reaped rewards which spread across into subsequent years. Some found the development task exceptionally stimulating and enjoyable. The initiative was also supported by the external examiners. However, the development did initially attract some hostility from internal staff, both academic and support. Some academic staff felt that their ownership of teaching materials was threatened, and were reluctant to offer ideas that could in theory subsequently be utilised by other staff. Several colleagues commented unfavourably on the 'front-end' loading of preparation time, and were anxious that they would not personally benefit from this in later years. One or two were initially unsympathetic to the educational objectives, and would have preferred to see equivalent financial investment in additional staff appointments. This attitude was also manifest in some reluctance to relinquish the control of the learning process to the students; occasionally, colleagues continued to deliver their 'capitalised' materials in the classroom, repeating verbally material which was readily available as text or on tape, rather than allowing students to study elsewhere to their own timetable.

From the students' perspective, the benefits were very considerable and regular formal student evaluation revealed initial enthusiasm from most, and year-on-year improvements in their responses. One notable phenomenon was that of 'familiarity'. As more modules began to diversify their style of delivery, students responded increasingly positively, perceiving the shift as a benefit rather than a threat to their anticipated level of achievement. Mature students were notably positive about the changes. Many students commented on and subsequently widened their study of particularly enjoyable elements. The only reluctance to engage with this came from students who had difficulty in organising their studies, and who somewhat resented the demand to take control of their own learning process. For some Higher Education students, attendance at a lecture session may be the full extent of the commitment that they are willing to make without protest.

Three years later, in 1995, the Higher Education Funding Council for England assessed the Department's geography provision, rating it as 'excellent'. In particular they noted:

"The Department provides an impressive range of high-quality learning materials, which have been effective in response to the challenge of rapidly expanding student numbers. Students were able to undertake independent study modules including self-paced exercises, and to experience student-led teaching… Desirable independent learning skills and a wide range of personal transferable skills are acquired, and many of these are demonstrated in the high quality of Level III dissertations" (HEFCE, 1995, p.4).

9.5.7   Conclusion

Overall, the project was very successful in meeting the objectives originally set. The student experience was enriched, and access arrangements benefited. Larger student groups were accommodated and the development undoubtedly addressed the need for students increasingly to be able to develop independent study skills. It also assisted the Department to manage dramatically increased student numbers. At the end of the project, the College decided to invest in a more permanent cross-College 'Flexible Learning Design and Development Unit' to assist other colleagues to make similar adjustments.

However, the pace of change in Higher Education in the UK during the 1990s highlights the need for careful planning of RBL initiatives. Not only colleagues, but courses come and go in response to changing patterns of undergraduate recruitment, and some of the initiatives taken proved to have relatively short shelf-lives. After two or three years of operation, the content may still be current, the approach appropriately challenging and stimulating for students, and the presentation attractive, but the course may no longer be needed by the host institution. This highlights the possibility of promoting wider inter-institutional approaches where the risks as well as the benefits of RBL approaches, can be shared.

Further information:

Carolyn Roberts, School of the Environment, Cheltenham and Gloucester College of Higher Education; email: crroberts@chelt.ac.uk

Activity 25

Review the case studies in this section and make a list of the ideas which a) you and b) your department would consider adopting. Prioritize the list and distinguish which ideas could be implemented for next semester/year and which would take longer to implement.

10 Guide to references and resources

The main references and resources have been referred to in the relevant sections. In this section I highlight some of the key references and resources.

10.1 General references on RBL

If you are seeking a general introduction to RBL have a look at: Brown & Smith (1996); Freeman & Lewis (1994); Gibbs *et al.* (1994a; 1994b); Lewis & Freeman (1994); Race (1993b); Rowntree (1990; 1992; 1994a; 1994b; 1997; 1998).

10.2 Examples of use of RBL in geography

If you wish to explore the use of RBL in geography look at the examples in the boxes and the case studies in Section 9. Some useful examples from the literature can be found in: Backler (1979); Clark & Gregory (1982); Church (1988); Cho (1982); Fox & Wilkinson (1977); Fox *et al.* (1987); Gold *et al.* (1991; 1996); Healey (1992); Healey & Ilbery (1993); Healey *et al.* (1996); Higgitt & Higgitt (1993a; 1993b; 1993c); Jenkins & Young (1983); Jones (1986); Keene (1982); Kemp & Goodchild (1991; 1992); Shepherd (1998).

10.3 RBL geography materials

A limited amount of RBL geography materials has been published. Most were developed by consortia including the ARGUS project (AAG, 1995) and the National Centre for Geographic Information Core Curriculum Project (Kemp & Goodchild, 1991; 1992). Some materials have been published by single institutions, such as the Open University (Brook, 1995) or small groups of colleagues (e.g. Kuby *et al.*, 1998). Further details of two consortia projects follow:

GeographyCal — The aims of the project are to specify, develop, test, and deliver a library of high quality transportable CAL modules to facilitate an efficient and effective teaching and learning environment for core topics, concepts and techniques in introductory undergraduate geography courses. The subjects of the modules were decided following a needs survey of all consortium departments. Discrete manageable subjects were chosen. Of the seventeen so far modules developed, six are in physical geography, six in human geography and five are concerned with techniques (Table 12). Each module provides one to three hours of student activity. A further seven modules are in production.

Table 12: *The GeographyCal modules*

Human Geography	Physical Geography
International economic change	Catchment systems (double)
Planning and development in the urban-rural fringe	Quaternary environmental change
International inequalities	Simulating slope development
Regional economic change	Biogeography and ecology
Environmental hazards	Weather and air quality
Social change at an international scale	Global tectonics
Geographical Techniques:	
Social survey design	GIS: Introductions and applications
Making sense of information	House hunting game
Map design	

Source: Healey, *et al*, 1998a

Further information:

Geoff Robinson and John Castleford, CTI Centre for Geography, Geology and Meteorology, University of Leicester and Mick Healey, Geography and Environmental Management Research Unit, Cheltenham and Gloucester College of Higher Education email: cti@le.ac.uk

http://www.geog.le.ac.uk/cti/Tltp; see also Box 39

Active Learning Modules on the Human Dimensions of Global Change — Ten modules were published in 1997:

- Introduction to the Human Dimensions of Global Change

- Living in the Biosphere: Production, Pattern and Diversity

- Human Driving Forces and their Impact on Land Use/Land Cover

- The Geography of Greenhouse Gas Emissions

- Population Growth, Energy Use, and Pollution: Understanding the driving forces of global change

- Global Change and Environmental Hazards: Is the world becoming more dangerous?

- Industry in Concert with the Environment: Technical change and industrial ecology as guiding forces for global environmental change

- Global Change and Urbanization in Latin America

- Think Locally, Act Globally! Linking local and global communities through democracy and environment

- Human Health in the Balance

About half of each module consists of activities that may be copied and distributed to students for in-class collaboration or independent assignments. The rest provides background information, supportive materials, copy-ready transparency masters, ideas for active-learning instructional approaches, and readings that can be used by the instructor or placed on library reserve for student use. At the time of writing each package cost US$16 (including shipping).

Further information:

AAG, 1710 Sixteenth Street NW, Washington DC 20009, USA

http://www.aag.org/HDGC/Hands_On.html

See also Box 11.

11 References

AAG (1995) *Activities and Readings in the Geography of the United States: Hands-on geography* (Washington DC: Association of American Geographers).

Adams, R., Gibbs, G., Jaques, D. & Watson, D. (1980) *Workbooks: A practical guide* (Oxford: Educational Methods Unit, Oxford Polytechnic).

Agnew, C. & Elton, L. (1998) *Lecturing in Geography* (Cheltenham: Geography Discipline Network, CGCHE).

Atherton, J. (1998) The Politics of RBL, paper posted on the Improving Student Learning Mailbase (9 March).

Backler, A. (1979) Mastery learning: a case study and implications for instruction, *Journal of Geography in Higher Education,* 3(1), pp.68-75.

Bainbridge, C. (1997) An interactive library skills workbook for engineering undergraduates, in: R. Hudson, S. Maslin-prothero & L. Oates (Eds.) *Flexible Learning in Action: Case studies in higher education*, pp.130-134 (London: Kogan Page).

Baum, T., Eggins, H., Jones, D. & Pointon, A.J. (1985) The fallacies of resource-based learning: a case for national coordination, *Higher Education Review*, 17(2), pp.53-59.

Baxter, E.P. (1990a) Comparing conventional and resource based education in chemical engineering: student perceptions of a teaching innovation, *Higher Education*, 19, pp.323-340.

Baxter, E.P. (1990b) Resource-based education in chemical engineering: the history and impact of a radical teaching innovation, *Studies in Higher Education*, 15, pp.223-240.

Bell, T. (1996) *Human Geography: People, places and change — Study Guide* (Upper Saddle River, NJ: Prentice Hall).

Birnie, J. & Mason O'Connor, K. (1998) *Practicals and Laboratory Work in Geography* (Cheltenham: Geography Discipline Network, CGCHE).

Bloom, B.S. (Ed.) (1956) *Taxonomy of Educational Objectives Handbook 1: Cognitive domain* (London: Longman).

Bradbeer, J. (1997) Society, nature and place: a final year core course in contemporary philosophical debates in geography, *Journal of Geography in Higher Education,* 21(3), pp.373-379.

Bradford, M. & O'Connell, C. (1998) *Assessment in Geography* (Cheltenham: Geography Discipline Network, CGCHE).

Brook C. (1995) *The Shape of the World: Explorations in Human Geography D215 Course Guide* (Milton Keynes: The Open University).

Brown, S. & Smith, B. (1996a) Introducing resources for learning, in: S. Brown. & B. Smith (Eds.) (1996) *Resource-Based Learning*, pp.1-9 (London: Kogan Page).

Cho, G. (1982) Experiences with a workbook for spatial data analysis, *Journal of Geography in Higher Education,* 6(2), pp.133-139.

Church, M. (1988) Problem orientation in physical geography teaching, *Journal of Geography in Higher Education,* 12(1), pp.52-65.

Clark, M.J. & Gregory, K.J. (1982) Physical geography techniques: a self-paced university course, *Journal of Geography in Higher Education,* 6(2), pp.123-131.

Clark, G. & Wareham, T. (1998) *Small-group Teaching in Geography* (Cheltenham: Geography Discipline Network, CGCHE).

Copyright Licensing Agency (1998) Higher Education Copying Accord, *Times Higher Education Supplement,* 17 April.

Cox, S. & Gibbs, G. (1994) *Course Design for Resource Based Learning: Social science* (Oxford: Oxford Brookes University, The Oxford Centre for Staff Development).

Cryer, P. & Elton, L. (1992) *Effective Learning and Teaching in Higher Education: 4. Active learning in large classes and with increasing student numbers* (Sheffield: CVCP Universities' Staff Development and Training Unit).

DES (Department for Education and Science) (1991) *Higher Education in the Polytechnics and College: Humanities and social sciences* (London: HMSO).

Digby, B. (1997) The geography of HIV/AIDS: an update, *Geofile* 309 (April).

Exley, K. & Gibbs, G. (1994) *Course Design for Resource Based Learning: Science* (Oxford: Oxford Brookes University, The Oxford Centre for Staff Development).

Farmer, D.W. & Mech, T.F. (1992) Editors' notes, in D.W. Farmer & T.F. Mech (Eds.) *Information Literacy: Developing students as independent learners*, pp.1-3 (San Francisco: Josey-Bass).

FDTL National Co-ordination Team (1998a) *FDTL Project Briefing 5: Developing teaching and learning materials* (Milton Keynes: Open University Centre for Higher Education Practice).

FDTL National Co-ordinating Team (1998b) *FDTL Project Briefing 7: Copyright* (Milton Keynes: Open University Centre for Higher Education Practice).

Fellows, S. J. (1994) Integration of open learning into mainstream higher education, in *University wide Change, Staff and Curriculum Development*, Staff and Educational Development Association Paper 83.

Fellows, S. (1997) Adopting a mixed-mode approach to teaching and learning: A case study of the University of Luton, in Hudson, R., Maslin-Prothero, S. & Oates, L. (Eds.) *Flexible Learning in Action: Case studies in higher education*, pp.147-152 (London: Kogan Page).

Foote, K. (1998) Building disciplinary collaborations in the worldwide web: Strategies and barriers, *Journal of Geography* (forthcoming).

Fox, M. & Wilkinson, T. (1977) A self-paced mastery instruction scheme in geography for a first year university course, *Journal of Geography in Higher Education*, 1(2), pp.61-70.

Fox, M.F., Rowsome, W.S. & Wilkinson, T.P. (1987) A decade of mastery learning: evolution and evaluation, *Journal of Geography in Higher Education*, 11(1), pp.3-10.

Freeman, R. & Lewis, R. (1995) *Writing Open Learning Materials: Staff development activities for FE and HE* (Lancaster: Framework Press Educational Publishers).

Funnell, D. & Browne, T. (1997) Using CAL and the WWW in geography methods teaching, *Learning Matters: Teaching and Learning at Sussex*, Issue 8. Also available at: http://www.sussex.ac.uk/Units/TLDU/LM/LMissue8/item7.html. A revised version has been accepted for publication in the *Journal of Geography in Higher Education*.

Gamson, A.W. & Chickering, Z.F. (1987) Seven principles for good practice in undergraduate education, *AAHE Bulletin*, 39(7), pp.3-7.

Gardiner, V. & D'Andrea, V. (1988) *Teaching and Learning Issues and Managing Educational Change in Geography* (Cheltenham: Geography Discipline Network, CGCHE).

Gibbs, G. (1992a) Control and independence, in: G. Gibbs & A. Jenkins (Eds.) *Teaching Large Classes in Higher Education: How to maintain quality with reduced resources*, pp.37-62 (London: Kogan Page).

Gibbs, G. (1992b) *Teaching More Students 5. Independent learning with more students* (Oxford: Oxford Centre for Staff Development, Oxford Brookes University).

Gibbs, G. & Jenkins, A. (Eds.) (1992a) *Teaching Large Classes in Higher Education: How to maintain quality with reduced resources* (London: Kogan Page).

Gibbs, G. & Jenkins, A. (1992b) An introduction: the context of changes in class size, in: G. Gibbs & A. Jenkins (Eds.) *Teaching Large Classes in Higher Education: How to maintain quality with reduced resources*, pp.11-22 (London: Kogan Page).

Gibbs *et al.* (1994a) *Course Design for Resource Based Learning* (Oxford: Oxford Brookes University, The Oxford Centre for Staff Development). A set of nine booklets each covering a different discipline or group of disciplines.

Gibbs, G., Pollard, N. & Farrell, J. (1994b) *Institutional Support for Resource Based Learning* (Oxford: Oxford Brookes University, The Oxford Centre for Staff Development).

Gold, J.R., Jenkins, A., Lee, R., Monk, J., Riley, J., Shepherd, I. & Unwin, D. (1991) *Teaching Geography in Higher Education: A manual of good practice* (Oxford: Basil Blackwell).

Gold, J.R., Revill, G. & Haigh, M.J. (1996) Interpreting the Dust Bowl: teaching environmental philosophy through film, *Journal of Geography in Higher Education,* 20(2), pp.209-221.

Haggett, P. (1983) *Geography: A modern synthesis* (New York: Harper and Row).

Haycock, C.A. (1991) Resource-based learning: a shift in the roles of teacher, learner, *NASSP Bulletin* (May), pp.15-22.

Healey, M. (1991) Teaching without lectures: the case of a final year course in industrial geography, in: Clark, G. (Ed.) *Geography and Enterprise*, pp.45-51 (Lancaster: Lancaster University and Institute of British Geographers Higher Education Study Group). Also available at http://www.chelt.ac.uk/gdn/ehe/ch7.htm

Healey, M. (1992) Curriculum development and 'enterprise': group work, resource-based learning and the incorporation of transferable skills into a first year practical course, *Journal of Geography in Higher Education,* 16(1), pp.7-19.

Healey, M. (1997) Geography and education: perspectives on quality in UK higher education, *Progress in Human Geography,* 21, pp.97-108.

Healey, M. (1998a) Developing and disseminating good educational practices: Lessons from geography in higher education, paper presented to The International Consortium for Educational Development in Higher Education's Second International Conference on 'Supporting Educational, Faculty & TA Development within Departments and Disciplines', Austin, Texas, 19-22 April 1998. Available at http://www.chelt.ac.uk/gdn/confpubl/iced.htm; forthcoming in Lewis, K (Ed.) *Conference Proceedings* (Austin: University of Texas at Austin).

Healey, M. (1998b) Developing and internationalising higher education networks in geography, editorial, *Journal of Geography in Higher Education*, 22(3), forthcoming.

Healey, M.J. & Ilbery, B.W. (1990) *Location & Change: Perspectives on economic geography* (Oxford: Oxford University Press).

Healey, M. & Ilbery, B. (1993) Teaching a course around a textbook, *Journal of Geography in Higher Education,* 17(2), pp.123-129.

Healey, M., Matthews H., Livingstone, I. & Foster I. (1996a) Learning in small groups in university geography courses: Designing a core module around group projects, *Journal of Geography in Higher Education,* 20(2), pp.167-180.

Healey, M., Robinson, G. & Castleford, J. (1996b) Innovation in geography teaching in higher education: developing the potential for computer-assisted learning, 28th International Geographical Congress, Commission on Geographical Education, *Innovation in Geographical Education: Proceedings*, pp.199-203 (Amsterdam, Centrum voor Educatieve Geografie Vrije Universiteit). Also available at http://www.chelt.ac.uk/gdn/confpubl/igu.htm

Healey, M., Robinson, G. & Castleford, J. (1998) Developing good educational practice: integrating *GeographyCal* into university courses, in: Bliss, E (Ed.) *Islands: Economy, Society and Environment: Proceedings of the Institute of Australian Geographers and New Zealand Geographical Society Second Joint Conference, University of Tasmania, Hobart 1997, New Zealand Geographical Society Conference Series* No. 19, pp.367-370 (Hamilton, New Zealand Geographical Society, University of Waikato). Also available at http://www.chelt.ac.uk/gdn/confpubl/cal.htm

Heathcote, D. & Pollard, N. (1994) Copyright, in Exley, K. & Gibbs, G. (1994) *Course Design for Resource Based Learning: Science,* pp.68-71 (Oxford: Oxford Brookes University, The Oxford Centre for Staff Development).

HEFCE (1995) *Quality Assessment Report Q85/96: Cheltenham and Gloucester College of Higher Education, Geography, January 1995* (Bristol: HEFCE).

Higgitt, M. & Higgitt, D. (1993a) *The Silverdale Peninsula Field Trail* (Lancaster: Department of Geography, Lancaster University).

Higgitt, M. & Higgitt, D. (1993b) *The Lane Estuary Field Trail* (Lancaster: Department of Geography, Lancaster University).

Higgitt, M. & Higgitt, D. (1993c) *The Lancaster Field Trail* (Lancaster: Department of Geography, Lancaster University).

Hodgson, B. (1993) *Key Terms and Issues in Open and Distance Learning* (London: Kogan Page).

Hunt, M. & Clark, A (1997) *A Guide to the Cost Effectiveness of Technology-Based Training* (Sheffield: DfEE).

Jenkins, A. & Smith, P. (1993) Expansion, efficiency and teaching quality: the experience of British geography departments 1986-91, *Transactions, Institute of British Geographers NS*, 18, pp.500-515.

Jenkins, A. & Youngs, M.J. (1983) Geographical education and film: an experimental course, *Journal of Geography in Higher Education,* 7(1), pp.33-44.

JISC Assist (1998) Copyright, *JISC Senior Management Briefing* 5, January.

Jones, A. (1986) Resource-based learning: shifting the load, *Journal of Geography in Higher Education,* 10(2), pp.159-168.

Jordan, S. & Yeomans, D. (1991) Whither independent learning? The politics of curricular and pedagogical change in a polytechnic department, *Studies in Higher Education*, 16, pp.291-308.

Keene, P. (1982) The examination of exposures of Pleistocene sediments in the field: a self-paced exercise, *Journal of Geography in Higher Education*, 6(2), pp.109-121.

Keller, F.S. (1968) Goodbye teacher, *Journal of Applied Behaviour Analysis*, 1, pp.79-89.

Kember, D. & Murphy, D. (1994) *53 Interesting Activities for Open Learning Courses* (Bristol: Technical and Educational Services).

Kemp, K.A. & Goodchild, M.F. (1991) Developing a curriculum in Geographic Information Systems: the National Centre for Geographic Information and Analysis Core Curriculum project, *Journal of Geography in Higher Education*, 15(2), pp.123-134.

Kemp, K.A. & Goodchild, M.F. (1992) Evaluating a major innovation in higher education: the NCGIA Core Curriculum in GIS, *Journal of Geography in Higher Education*, 16(1), pp.22-35.

Kolb, I.D. (1984) *Experiential Learning: Experience as the source of learning and development* (New Jersey: Prentice Hall).

Kuby, M., Harner, J. & Gober, P. (1998) *Human Geography in Action* (Chichester: John Wiley).

Laurillard, D. (1993) *Rethinking University Teaching: A framework for the effective use of educational technology* (London: Routledge).

Lewis, R. & Freeman, R. (1994) *Open Learning in Further and Higher Education: A staff development programme* (Lancaster: Framework Press Educational Publishers).

Library Association (1996) *Copyright in Further and Higher Education Libraries* (London: Library Association)

Livingstone, I., Matthews, H. & Castley, A. (1998) *Fieldwork and dissertations in Geography* (Cheltenham: Geography Discipline Network, CGCHE).

Massey, D. (1995) Geographical imaginations, in J. ALLEN & D. MASSEY (Eds.) *Geographical Worlds*, pp.5-51 (Oxford: Oxford University Press in association with The Open University).

McCracken, R. & Gilbart, M. (1995) *Buying and Clearing Rights: Print, broadcast and multimedia* (London: Blueprint).

McGhee, P. (1997) Flexible learning as a management issue, in: R. Hudson, S. Maslin-Prothero & L. Oates (Eds.) *Flexible Learning in Action: Case studies in higher education*, pp.160-164 (London: Kogan Page).

McHenry, K.E., Stewart, J.T. & Wu, J.L. (1992) Teaching resource-based learning and diversity, in D.W. Farmer & T.F. Mech (Eds.) *Information Literacy: Developing students as independent learners*, pp.55-62 (San Francisco: Josey-Bass).

McKeachie, W.J. (1994) *Teaching tips: Strategies, research, and theory for college and university teachers* (Lexington: Massachusetts, D.C. Heath).

McKendrick, J.H. & Bowden, A. (1997) Audio Visual Learning Resources in Geography: UK survey, mimeo giving basic survey results, available from John McKendrick, Glasgow Caledonian University.

Meyers, C. & Jones, T.B. (1993) *Promoting active learning* (San Francisco: Josey-Bass).

Moore, K. (1998) Notes from a virtual field course, *CTI Centre for Geography, Geology and Meteorology GeoCal*, 18, pp.20-21.

Moser, S. & Hanson, S. (1996) *Developing Active Learning Modules on the Human Dimensions of Global Change: Notes on Active Pedagogy: A supplement to the active learning modules* (Washington: Association for American Geographers). Also available at: http://www.utexas.edu/depts/grg/virtdept/library/activeped/activeped.html

NCIHE (The National Committee of Inquiry into Higher Education) (1997). *Higher Education in the Learning Society* (Norwich: HMSO).

Noble, P. (1980) *Resource Based Education in Post Compulsory Education* (London: Kogan Page).

OCSD (The Oxford Centre for Staff Development) (various) *Teaching Large Classes series* (Oxford: Oxford Brookes University, The Oxford Centre for Staff Development).

O'Hagan, C.M. (1995) Custom video for flexible learning, *Innovations in Education and Training International*, 32(2), pp.131-138.

Race, P. (1992) *53 Interesting Ways to Write Open Learning Materials* (Bristol: Technical and Educational Services).

Race, P. (1993a) *Never Mind the Teaching Feel the Learning* (Birmingham: Staff and Educational Development Association, SEDA Paper 80).

Race, P. (1993b) *The Open Learning Handbook: Promoting quality in designing and delivering flexible learning* (London: Kogan Page).

Race, P. (1996) Helping students to learn from resources, in S. Brown. & B. Smith (Eds.) *Resource-Based Learning*, pp.22-37 (London: Kogan Page).

Race, P. (1997) A fresh look at independent learning, at http://www.lgu.ac.uk/deliberations/eff.learning/indep.html 12 January 1997.

Ramsden, P. (1992) *Learning to Teach in Higher Education* (London: Routledge).

Raymond, A. (1998) *Copyright Made Easier* (London: Aslib).

Roberts, C. (1995) *Costing innovation: staff development, resources and delivery*, paper presented to Geography in Education Network for Empowerment workshop on Empowerment in Teaching Geography, 11 July, Northampton.

Robinson, G., Castleford, J. & Healey, M. (1998) Consorting, collaborating and computing: the *GeographyCal* project, in: Bliss, E (Ed.) *Islands: Economy, Society and Environment: Proceedings of the Institute of Australian Geographers and New Zealand Geographical Society Second Joint Conference, University of Tasmania, Hobart 1997, New Zealand Geographical Society Conference Series* No. 19, pp.370-374 (Hamilton: New Zealand Geographical Society, University of Waikato).

Robinson, J. (1997) Staff development for flexible learning at Stockport College, in: R. Hudson, S. Maslin-Prothero & L. Oates (Eds.) *Flexible Learning in Action: Case studies in higher education*, pp.175-179 (London: Kogan Page).

Rolls, D. & Watts, S. (1994) Study packages used mainly to replace part of first-year lecture programme in a modular degree course, *Journal of Geography in Higher Education*, 18(2), p.249.

Rowntree, D. (1990) *Teaching Through Self-Instruction: How to develop open learning materials* (London: Kogan Page).

Rowntree, D. (1992) *Exploring Open and Distance Learning* (London: Kogan Page).

Rowntree, D. (1994a) *Preparing Materials for Open, Distance and Flexible Learning: An action guide for teachers and trainers* (London: Kogan Page).

Rowntree, D. (1994b) *Teaching with Audio in Open and Distance Learning* (London: Kogan Page).

Rowntree, D. (1997) *Making materials-Based Learning Work: Principles, politics and practicalities* (London: Kogan Page).

Rowntree, D. (1998) Motivating teachers for materials-based learning, *The International Journal for Academic Development,* 3(1), pp.47-53.

Shepherd, I.D.F. (1998) *Teaching and Learning Geography with Information and Communication Technologies* (Cheltenham: Geography Discipline Network, CGCHE).

Shepherd, I.D.F. & Bleasdale, S. (1993) Student reading and course readers in geography, *Journal of Geography in Higher Education*, 17(2), pp.103-121.

Sheridan, A.M. (1987) *Human Geography: An introduction* (State College, Pennsylvania: Department of Geography and Department of Independent Learning Penn State University).

Skinner, E. (1997) Achieving quality in Local Policy, *Journal of Learning and Teaching*, 3(1), pp.11-15 (Cheltenham: Cheltenham and Gloucester College of Higher Education).

Tierney, J. (1992) Information literacy and a college library: a continuing experiment, in D.W. Farmer & T.F. Mech (Eds.) *Information Literacy: Developing students as independent learners*, pp.63-71 (San Francisco: Josey-Bass).

University of Plymouth (1996) *Copyright: Some guidelines for university staff* (Plymouth: University of Plymouth).

Unwin, D. (1998) Virtual helicopters fly through time and space, *Times Higher Education Supplement*, 26 June.

Vujakovic, P., Livingstone, I & Mills, C. (1994) Why work in groups? *Journal of Geography in Higher Education,* 18(1), pp.124-127.

Whitelegg, J. (1982) The use of self-produced video material in first-year undergraduate practical classes, *Journal of Geography in Higher Education,* 6(1), pp.21-28.

12 Acknowledgements

A large number of people have helped in the production of this Guide. It would be invidious to name any particular individuals. Most importantly have been the many colleagues who contributed details about their experiences of developing and using RBL in geography. Some of the details were culled from the literature, but the majority were obtained through direct contact. The contact names for those I have managed to include are given in the boxes and the case studies in Section 9. I have also received a lot of support and encouragement from the members of the GDN Team and the Project Advisers, many of whom commented on drafts of the Guide. I have tried to respond to most of their suggestions, but they cannot be held responsible either for what I have said or what I have excluded.

I have attempted to trace the Copyright holders for permission to use various materials. I apologise if any have inadvertently been omitted. The following copyright holders have given permission:

- The Oxford Centre for Staff and Learning Development, Oxford Brookes University for the reproduction of Gibbs *et al.* (1994, pp.28-29) in Section 6.4; and for permission to update Exley & Gibbs (1994, pp.53-56) in Section 9.1 and Cox & Gibbs (1994, pp.56-58) in Section 9.4.

- Framework Press Educational Publishers, c/o Folens Publishers Limited, for the reproduction of Freeman & Lewis (1995, p.51) in Table 8.

- Kogan Page Ltd for the reproduction of Rowntree (1994a, pp.103-104 and 139-140) in Tables 6 and 7; and Race (1996, p.34) in Table 3.

- The Copyright Clearance Agency for the reproduction of CLA (1998) in Figure 8 (Appendix II).

- Carolyn Roberts, Cheltenham & Gloucester College of Higher Education for the reproduction of Figure 2.

Appendix I
Comments on activities

Most of the Activities in this Guide are open ended and have been designed so that you can reflect on the relevance of the material discussed in the text to your situation. Readers will have different interests and backgrounds and hence it would be difficult to provide helpful commentary to these. However, there are a few activities where some commentary may be useful.

Activity 4

The first and third definitions were produced by librarians, the second and fourth definitions were written by educational developers:

1. Farmer & Mech, 1992, p.1

2. Race, 1997

3. McHenry *et al.*, 1992, p.55

4. Gibbs *et al.*, 1994a, p.5

Activity 10

Race (1993a) asked a large number of people about the processes by which they became good at something and the causes of unsuccessful learning experiences. The most frequent answers to these questions are along the following lines:

Q1   Practice; doing it; trial and error; getting it wrong at first and learning from mistakes.

Q2   Reactions of other people; feedback; compliments; seeing the results.

Q3   Lack of opportunity to practise or to learn safely from mistakes; 'bad' feedback — critical feedback given in a hostile or negative way; no motivation; fear of failure; couldn't see why it was worth doing; lack of time to make sense of it; unable to understand it before moving on.

Interestingly, relatively few people answered Q1: 'by being taught'. Analysis of the answers Race received to these questions led him to identify four key factors influencing learning, which he labelled wanting, doing, feedback and digesting (Table 2).

Appendix II
Copyright

"Whilst there is a general recognition that potentially copyright is an important topic, many in higher education institutions (HEI) avoid the subject on the grounds that it is perhaps too complex to worry about"

(JISC, 1998)

"However, the law of copyright is relatively straightforward, and we are assured that, so long as we act in good faith and follow a few simple guidelines, no one is likely to end up in court and far less behind bars!"

(FDTL National Co-ordination Team, 1998b, p.1)

In this Appendix I try to provide you with guidance on some of the copyright issues you will face in copying learning materials for use by your students. The content is largely based on material in the sources cited later. It is structured around a set of questions.[*]

What does this Guide cover?

Copyright is a complex and evolving subject and this guidance is necessarily limited and partial. For a start I shall concentrate on written materials. I shall also limit myself to the situation in the UK. Although I have attempted to check the accuracy of what I have written, the usual caveat applies, namely that it does not constitute legal advice and it is your responsibility to check that you are staying within the copyright law! In this context it would be sensible to consult with the your institution's copyright officer (probably in the Library/ Learning Centre) if you have one, before say designing your module around a course reader or a set of materials provided as a series of handouts for your students. Many institutions are currently preparing their own internal guidance interpretation of the latest The Copyright Licensing Agency (CLA) agreement.

What is copyright?

Broadly speaking copyright is the protection given by law (The Copyright, Designs and Patents Act 1988) to an author for his/her work. The word 'author' means not only writers but also artists, musicians, computer software writers and so on. Copyright protection is automatic and does not need to be registered. However, it expires, most commonly 70 years after the death of the author.

Who holds copyright and how can permission be obtained?

Generally speaking, copyright owners have the sole right to use their work or to authorize others to use them, although there is a 'fair dealing' provision which allows an individual to make a copy for research or private study (see below). Copyright can be transferred from

---

[*] *Thanks are due to Richard McCracken (Rights Manager, Open University Rights Department) and Lyn Oates (Head of Learning and Teaching Support, Cheltenham and Gloucester College of Higher Education) for their helpful comments on an earlier draft of this Appendix.*

person to person by gift, sale, licence or on death. As a result finding the person who owns or controls the rights may sometimes be time consuming.

What can I (and my students) copy for private research/study?

Under the 'fair dealings' provision of the Act, an individual may make a copy for research or private study. Although there is no definition of 'fair dealing' it is normally accepted that the limits should be:

- one article from any issue of a journal;

- one chapter or up to 5% of a whole book or pamphlet or report.

What can I copy for use in my teaching?

The CLA has recently agreed a *Higher Education Copying Accord* with the Committee of Vice-Chancellors and Principals which provides a blanket licence for the photocopying of copyright material within the HE sector (CLA, 1998).

The CLA blanket licence for HE allows photocopying of a wide range of copyright material without seeking clearance within the following general limitations:

- Not more than one chapter from a book, or one article from a journal/periodical;

- No more than 5% of a given work, whichever is the greater.

The cost of the licence to HEIs is based on a discounted page rate of 5 pence, and assumes a notional level of copying per student per annum of only 65 pages, a far smaller number than actually is or now may be legitimately copied.

At first sight the Accord seems to allow photocopying of material for use in class.

"Where the lecturer or tutor wishes to distribute copyright materials to students on a genuinely ad hoc basis during a particular course of study, he/she may do so subject to those limitations, under the blanket licence. Provided that the copied materials do not constitute a course pack and provided also that the limitations are kept to, the lecturer or tutor may copy as many pages as he considers are reasonably necessary for the course of study" (CLA, 1998).

However, the definition of a course pack is such that this allowance is useful in a limited number of circumstances.

"A course pack is a compilation of materials (whether bound or loose leaf) of four or more photocopied extracts from one or more sources, totalling over 25 pages of copyright material, designed to support a module or course of study, irrespective of whether the materials are copied at the start of the course, or at intervals during the course, or are placed in the short loan reserve or equivalent for systematic copying by students at intervals throughout the course. All course packs must be cleared through CLA's rapid clearance service (CLARCS) but the additional copyright fees and the cost of course pack production can be passed on to students" (CLA, 1998).

An example of *ad hoc* photocopying which would seem to be allowable might be where each student/group receives a different photocopy, perhaps different case studies for use in a workshop exercise.

How do I obtain clearance for course packs?

In some HEIs this would normally be done by your institutions' copyright officer or someone to whom they have delegated this task, but a variety of arrangements exist and in some institutions it is the responsibility of the course team (Figure 8). The CLA (1998) say:

"To obtain clearance, or for the details of course pack copying fees, please contact the CLA by telephone on 0171-436 5931, fax: 0171-436 3986, or email: cla@cla.co.uk. Having initially registered with the CLA as an authorised account-holder, please enter onto permission request forms full bibliographic details of the entire course pack contents, including the number of pages copied from each work, and the number of students (as lawful users) registered for the course. CLA will issue an authorisation number of students (as lawful users) registered for the course. CLA will issue an authorisation number promptly, stating the required fair royalties due to the copyright holders on licensees' behalf to request clearance where CLA itself is not mandated to clear specific items" (CLA, 1998).

This is the CLA recommended practice, but many institutions have their own distinct local arrangements for obtaining copyright clearance.

Can photocopies be put in the Short Loan collection in the library?

Yes, the CLA licence allows copies to be made to be put in the library, subject to the normal conditions of the licence. Photocopies of copyright material may be made without the explicit consent of the copyright holder for inclusion in short loan collections:

- from original published editions already held in stock;

- or under the terms of a CLA licence;

- or by acquiring copyright-fee paid copies from the British Library Document Supply Centre. (Photocopies obtained via the Inter-library loan service on the basis that they are for research or private study cannot subsequently be put into the Short Loans collection. A higher fee needs to be paid at the time of ordering to cover the copyright fee).

The CLA licence permits photocopying of items held in short loan collections obtained in these ways.

Short loan collections cannot, however, be used to circumvent the provisions of the licence with regard to course packs.

"The inclusion of items at the request of academic staff which are the object of instruction or encouragement to students to create sets of copies for each person studying a module or unit, is considered to constitute creation of course packs without permission. It is therefore a breach of the licence" (CLA, 1998).

Again there are different local practices. Many university libraries and department resource centres have destroyed, or are in the process of removing, items from short loan collections for which they cannot prove a legal source. For example, this was the main reason that we had to close the geography and geology collection of articles at Cheltenham and Gloucester College of Higher Education. Many of the articles were copies of copies and in most cases it was not possible to identify which articles met the CLA requirements and which did not. Libraries and resource centres also need to devise useful ways of filing their photocopy collections that does not constitute making them into course resource packs!

Figure 8: *Flowchart guide to course pack clearance (Source: CLA, 1998)*

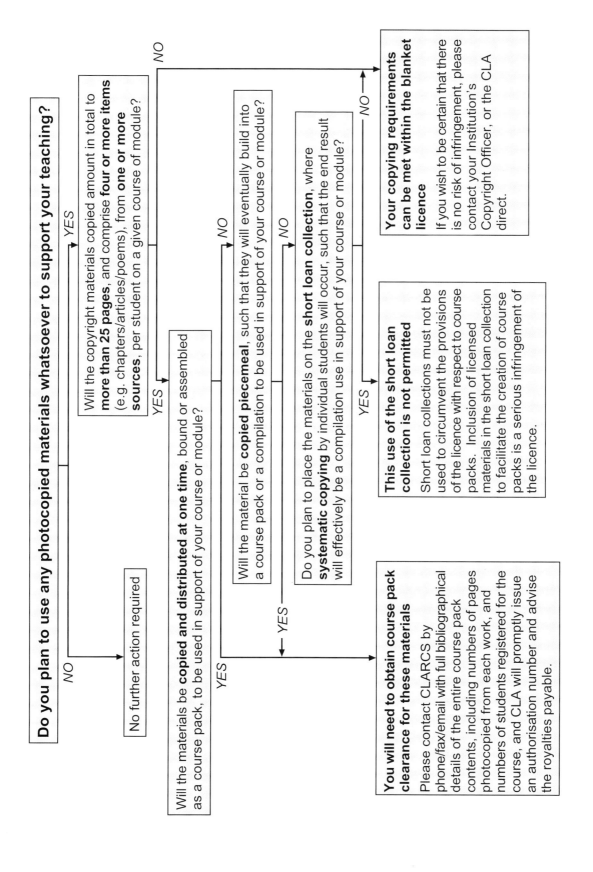

Can I make copies of maps?

Most HEIs have a separate licence from the Ordnance Survey which allows up to four copies of any Ordnance Survey map to be copied; extracts should not exceed A4 size at scale 1:25000 and smaller, and A3 at all scales larger than 1:25000. HEIs may also negotiate a licence with The British Geological Survey for limited photocopying of their maps.

What further sources on copyright might I find useful?

There are many books, articles and guides which provide more detail on copyright issues (e.g. Copyright Licensing Agency, 1998; FDTL National Co-ordination Team, 1998b; Heathcote & Pollard, 1994; Library Association, 1996; McCracken & Gilbart, 1995; Raymond, 1998; University of Plymouth, 1996). There are also some useful Web sites. For example, The National Centre for Legal Education's Web site (http://www.law.warwick.ac.uk/ncle/) provides links to information on copyright; while the Joint Information Systems Committee (JISC) has published a series of Guidance Papers on copyright, dealing particularly with the rapidly changing situation affecting electronic materials (http://www.jisc.ac.uk/pub/copyright/start.htm).